Micro- and Nano-Scale Sensors and Transducers

Micro- and Nano-Scale Sensors and Transducers

Ezzat G. Bakhoum

CRC Press
Taylor & Francis Group
Boca Raton London New York

CRC Press is an imprint of the
Taylor & Francis Group, an **informa** business

CRC Press
Taylor & Francis Group
6000 Broken Sound Parkway NW, Suite 300
Boca Raton, FL 33487-2742

First issued in paperback 2019

© 2015 by Taylor & Francis Group, LLC
CRC Press is an imprint of Taylor & Francis Group, an Informa business

No claim to original U.S. Government works

ISBN-13: 978-1-4822-5090-9 (hbk)
ISBN-13: 978-1-138-89430-3 (pbk)

Visit the Taylor & Francis Web site at
http://www.taylorandfrancis.com

and the CRC Press Web site at
http://www.crcpress.com

Contents

Preface

The past decade has witnessed substantial advances in sensor technology. The emerging new field of nanotechnology and nanofabrication has allowed the creation of a new array of sensors and transducers with remarkable properties. Pressure sensors, gas sensors, optical sensors, biological sensors, etc., have all seen dramatic improvements in characteristics such as sensitivity and dynamic range, in addition to substantial miniaturization. This book presents a summary of the state of the art of sensor and transducer technology as of 2014. Although a very large number of books on nanotechnology in general currently exist in the marketplace, recent advances in micro- and nano-scale sensors and transducers are not adequately represented in the literature. This book attempts to fill that gap. The intended audience for this book is practicing industry engineers, corporate and government researchers, and graduate students in electrical engineering, mechanical engineering, and physics.

The main topics covered in the book are the following:

- Pressure Sensors (Chapter 1): The first chapter presents the novel new structures of pressure sensors, used extensively in such applications as mechanical pressure sensing, gas pressure sensing, atmospheric pressure sensing, etc. Pressure sensors that are based on capacitance variation, in particular, have benefited from the recent advances in nanotechnology and nanofabrication, and this type of pressure sensor is covered extensively.

- Motion and Acceleration Sensors (Chapter 2): Motion and acceleration sensors are used in many applications, from automobile air bags to projectiles to smart tablets and cell phones. This category of sensors has also benefited greatly from the nanotechnology/nanofabrication revolution. The novel structures of the new motion and acceleration sensors that appeared recently in archival publications along with their amazing characteristics are presented in Chapter 2.

- Gas and Smoke Sensors (Chapter 3): Highly sensitive and miniature gas and smoke sensors that are based on nanostructured electrodes have been introduced recently in the literature. Chapter 3 describes these sensors.

■ Moisture Sensors (Chapter 4): Novel new techniques based on nanotechnology for detecting atmospheric moisture as well as moisture inside small electronic components have also appeared in the literature recently. Although not yet available commercially, these anticipated new sensors are ultraminiature in size yet ultrasensitive. Chapter 4 introduces these sensors.

■ Optoelectronic and Photonic Sensors (Chapter 5): Nanotechnology has revolutionized a number of classical applications by allowing the integration of optical sensing techniques into such applications. Advanced new products in this category include optical microphones, fingerprint readers, and highly sensitive seismic sensors. These advanced new applications are covered in Chapter 5.

■ Biological Sensors, Chemical Sensors, and the so-called "Lab-on-a-Chip" (Chapter 6): Another important revolution based on nanotechnology has culminated in multipurpose biological and chemical analysis devices where each device is fully contained in one integrated circuit (the so-called Lab-on-a-Chip) in addition to other advanced chemical and biological sensors. A survey of these sensors is given in Chapter 6.

■ Electric, Magnetic, and RF/Microwave Sensors (Chapter 7): Enormous advances in electric field, magnetic field, and RF/Microwave sensors, driven by nanotechnology, have occurred recently. A description of these sensors, along with their applications, is given in Chapter 7.

■ Integrated Sensor/Actuator Units and Special Purpose Sensors (Chapter 8): The last chapter of the book is dedicated to integrated sensor/actuator units and special-purpose sensors. New devices that benefited from nanotechnology, such as new icing detectors for aircraft, new microfluidic sensor/actuator units for microrobots and inkjet printers, etc., are described in Chapter 8.

With the information provided in this book, the corporate researcher or design engineer will be able to:

■ understand the differences between the new sensor and transducer technology (which is mainly based on nanotechnology and nanofabrication) and the older or "classical" sensor technologies;

■ make an informed selection of a sensor or transducer for a particular application;

■ become knowledgeable about the technologies that are available commercially at the present time and the technologies that are anticipated to become available within a time span of a few months to a few years.

Each chapter of the book ends with a set of quizzes/short questions, along with answers.

For MATLAB® and Simulink® product information, please contact:
The MathWorks, Inc.
3 Apple Hill Drive
Natick, MA, 01760-2098 USA
Tel: 508-647-7000
Fax: 508-647-7001
E-mail: info@mathworks.com
Web: www.mathworks.com

Chapter 1

Pressure Sensors

1.1 Capacitive Pressure Sensors

Among the pressure sensors that are widely used in the industry, capacitive pressure sensors are particularly noteworthy. These sensors are characterized by very low temperature hysteresis and pressure hysteresis, in addition to low power consumption [1–8]. Traditional capacitive pressure sensors, however, suffer from inherently poor resolution (a typical capacitive pressure sensor offers a total change in capacitance of only a few pico-farads, which usually necessitates the use of a sophisticated interface/compensation circuit to sense the very small variations in capacitance). New capacitive pressure sensors with extremely high resolution and sensitivity, based on nanotechnology, were introduced recently [9]. This section introduces the mercury droplet capacitive pressure sensor, the sensor with the highest reported sensitivity and resolution. This type of sensor is currently in production and should be commercially available in early 2015.

1.1.1 Structure

The recently introduced mercury-droplet capacitive pressure sensor has demonstrated a change in capacitance of approximately 6.73 μF over a pressure range of 0 to 3 kPa. The sensitivity of this type of sensor is therefore 2.24 μF/kPa, substantially higher than any of the known types of capacitive pressure sensors. The basic concept of the new sensor is to mechanically deform a drop of mercury that is separated from a flat aluminum electrode by a very thin layer of a dielectric material, so as to form a parallel-plate capacitor where the electrode area is variable to a high degree. This principle is illustrated in Figure 1.1 below.

The principle of the new device, therefore, is to create a capacitor with a variable electrode area, rather than a variable interelectrode spacing as commonly done in the

1

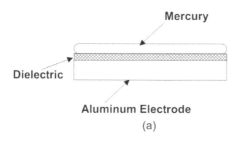

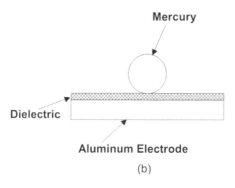

Figure 1.1: (a) A drop of mercury is flattened against an aluminum electrode that is covered with a layer of a dielectric material. A parallel-plate capacitor with one liquid electrode is formed. (b) Under zero pressure, the mercury drop returns to its nearly spherical shape. The change in capacitance between the two configurations (which is proportional to the change in the contact area of the liquid electrode) can be several hundred fold.

devices shown in the literature. The detailed structure of the new sensor, together with the test data, is given in the following sections. Table 1.1 below lists the four most important parameters of the new sensor: sensitivity, linearity, pressure hysteresis, and temperature hysteresis, as compared to the other known types of pressure sensors.

As the table shows, the sensitivity of the new sensor is substantially higher than any of the known types of pressure sensors. The hysteresis error is also substantially lower than that of other sensors. The drawback, however, is that the maximum temperature-related error is slightly worse than that of the other capacitive pressure sensors (due to the thermal expansion of the mercury droplet, particularly at high temperatures), although it is still better than the temperature-related error offered by piezoresistive sensors. Another important fact to mention is that while the sensor is nonlinear (like most other capacitive sensors), the equation that relates the capacitance to the applied pressure is exactly known, as will be demonstrated in the following sections.

The basic structure of the new sensor is shown in Figure 1.2. A drop of mercury of a 3 mm diameter is placed on top of a flat aluminum electrode that is covered with a

Table 1.1   Comparison of the new sensor to the two most important types of pressure sensors: Capacitive and piezo-resistive (see [10–14]).

	Sensitivity	Linearity	Pressure Hysteresis	Temperature Hysteresis (For temp. range of −10°C to +80°C)
Piezoresistive pressure sensors	Up to 25 mv/kPa	Generally linear	Up to ± 1% FSO	Up to ± 2% FSO
Capacitive pressure sensors	Up to 0.2 nF/kPa	Generally nonlinear	Up to ± 0.1% FSO	Up to ± 0.5% FSO
New sensor (uncompensated)	2.24 µF/kPa	Nonlinear	Less than ± 0.05% FSO	Up to ± 1.5% FSO

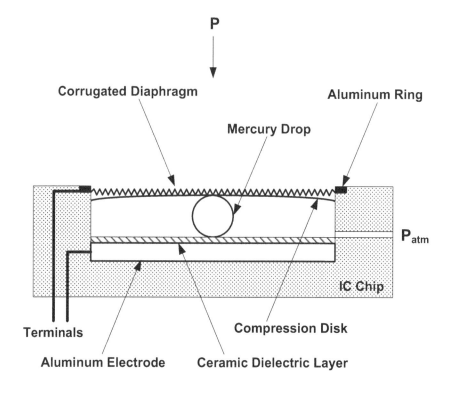

Figure 1.2: Mechanical structure of the sensor.

1 μm thick layer of a ceramic material that has a very high dielectric constant (specifically, $BaSrTiO_3$, with a dielectric constant of 12,000–15,000). The drop is held in place by means of an aluminum disk that serves as the compression mechanism. The compression disk, in turn, is acted upon by means of a corrugated stainless steel diaphragm, as shown (those corrugated diaphragms are available from a number of industrial suppliers). The compression disk is given a slight curvature, as shown in the figure, such that the spacing between the disk and the ceramic layer is exactly 3 mm at the center, but less than 3 mm everywhere else. In this manner, the mercury drop will be forced to the center each time the stainless steel diaphragm retracts. The diaphragm is held in place by means of a thin aluminum ring, as shown (conductive paste between the rim of the diaphragm and the ring allows an air-tight seal to be formed). The entire assembly is mounted inside an open-cavity, 24-pin DIP IC package. A photograph of the components of the sensor is shown in Figure 1.3.

Since the air that surrounds the mercury droplet must be allowed to exit from the sensor and re-enter as the sensor is pressurized/depressurized, an atmospheric pressure relief conduit is drilled in the IC package, as shown on the right hand side of Figure 1.2. In most applications, that conduit will be connected to an atmospheric pressure environment via, for example, an external tube to be connected to the sensor

Figure 1.3: Components of the sensor. The sensor is totally mounted inside a standard 24-pin DIP IC package (dimensions: 30 mm × 14 mm).

(it will be advantageous to connect the pressure relief conduit to the ambient environment through a moisture isolation chamber, in order to prevent moisture from penetrating inside the sensor). In applications where it is desired to detect pressures that are lower than the atmospheric pressure at sea level (like aircraft altitude applications, for example), then a suitable vacuum can be initially applied to the pressure relief conduit (in which case the mercury drop will be initially flattened at sea level).

Concerning the 1 μm thick layer of $BaSrTiO_3$, it is deposited on the surface of the aluminum electrode by using the electrophoretic deposition technique [15]. Figure 1.4 shows a scanning electron microscope (SEM) picture of the ceramic layer deposited on the surface of the electrode. The dielectric constant of the ceramic layer was found to be approximately 12,000, as expected for this material [16, 17].

A word is now in order concerning the interface circuit used with the sensor. At the present time, the interface circuit used is a 555 timer working in an oscillator mode, essentially for converting the capacitance to frequency. Such a circuit is very well known in the literature and is described in references such as [18]. The equation that characterizes the 555 oscillator is $(1 - \exp[-1/2fRC]) = 2/3$ [18]. Given a known resistance R, the value of the unknown capacitance C can be easily calculated from that equation by observing the frequency f of the resulting square wave. The miniature, surface-mount 555 chip is integrated inside the open cavity package shown in Figure 1.3 (the chip is mounted underneath the sensor and is not shown in the photograph). It is to be pointed out that the interface circuit does not amplify or

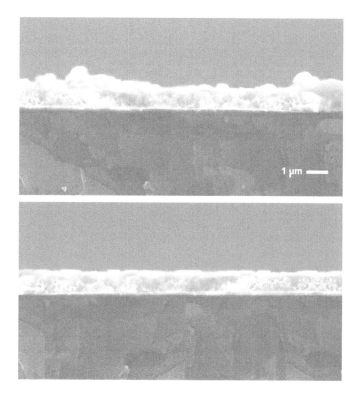

Figure 1.4: An SEM micrograph of the 1 μm thick layer of BaSrTiO$_3$ deposited on the surface of the aluminum electrode (upper) and a micrograph of the same layer after fine polishing (lower).

compensate for the nonlinear characteristics of the sensor (i.e., the sensor is uncompensated, as shown in Table 1.1). The signal-to-noise ratio of the entire assembly (sensor + interface circuit) was found to be substantially high (40 dB or higher).

1.1.2 Theory

Figure 1.5 shows the geometry of a drop of mercury that is deformed between two solid surfaces. The vertical pressure that is acting on the drop is P, and the lateral pressure is the atmospheric pressure P_{atm}. P_{int} is the internal pressure, and R is the radius of curvature of the part of the surface of the liquid that is not flattened, as shown.

The internal pressure P_{int} in the liquid must be balanced by the atmospheric pressure plus the Laplace pressure, or the pressure due to surface tension [19], that is,

$$P_{int} = P_{atm} + \frac{2\gamma}{R} \qquad (1.1)$$

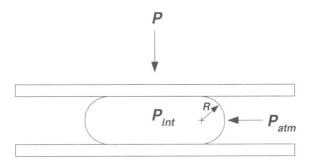

Figure 1.5: Pressures and geometry in the deformation of a drop of mercury.

where $2\gamma/R$ is the Laplace pressure and γ is the surface tension of mercury. As the drop of mercury is flattened, the difference in the internal pressure will be equal to the applied pressure P, i.e.,

$$P = P_{int} - (P_{int})_0 \tag{1.2}$$

where $(P_{int})_0$ is the internal pressure at zero applied pressure. By using Eq. (1.1), Eq. (1.2) can be written as follows:

$$P = P_{atm} + \frac{2\gamma}{R} - \left(P_{atm} + \frac{2\gamma}{R_0} \right) \tag{1.3}$$

where R_0 is the original unflattened radius of the drop. Hence,

$$P = 2\gamma \left(\frac{1}{R} - \frac{1}{R_0} \right) \tag{1.4}$$

In the sensor described here, the total pressure that is applied to the device must be equal to the pressure P plus an additional pressure that is needed to deform the metal diaphragm. That additional pressure was determined experimentally and is discussed further below. The capacitance of a parallel plate capacitor is now given by the well-known equation [24]

$$C = \frac{\varepsilon A}{d} \tag{1.5}$$

where ε is the permittivity of the dielectric medium, A is the surface area of the electrodes, and d is the thickness of the dielectric. It is to be noted from Figure 1.2 that the contact area, or "wetting" area, A, of the undeformed drop of mercury is very small and that an additional capacitance (beside the one defined by Eq. (1.5)) exists between the metal sphere and the flat electrode, with air being the dielectric. Essentially, that parasitic capacitance can be calculated from theoretical models given in references such as [20]. That parasitic capacitance, however, is totally negligible in comparison with the main capacitance given by Eq. (1.5), due to the very high dielectric constant of the ceramic dielectric layer that exists between the electrodes (that is, the air capacitance is negligible in comparison with the capacitance of the ceramic). Equation (1.5) is therefore the capacitance equation that is relied upon in the present work.

The wetting area A of the deformed drop (see Figure 1.5) can be calculated from the fact that the volume of the droplet remains constant. The volume of the undeformed droplet is equal to $4\pi R_0^3/3$, where R_0 is the radius of the undeformed droplet. Calculating the wetting area of the deformed droplet is a simple but rather lengthy and uninformative exercise in geometry. We just give the result:

$$A = \pi \left[\sqrt{\frac{\pi^2 R^2}{16} + \frac{2R_0^3}{3R}} - \frac{\pi}{4}R \right]^2 \tag{1.6}$$

where R is the radius of curvature of the part of the surface of the liquid that is not flattened (see Figure 1.5). The capacitance C between the two electrodes can then be determined from Eq. (1.5). In the device considered here, the capacitance is actually measured, and then other parameters such as radius of curvature and, finally, pressure, can be calculated from the equations. Equations (1.5) and (1.6) can be solved to give the radius of curvature R as a function of the capacitance C. This relationship is found to be

$$R = \sqrt{\frac{Cd}{\varepsilon\pi^3} + \frac{4}{3}R_0^3\left(\frac{\varepsilon}{\pi Cd}\right)^{1/2}} - \sqrt{\frac{Cd}{\varepsilon\pi^3}} \tag{1.7}$$

From Eqs. (1.4) and (1.7), the pressure P acting on the mercury drop can be formulated as a function of capacitance:

$$P = 2\gamma \left/ \left(\sqrt{\frac{Cd}{\varepsilon\pi^3} + \frac{4}{3}R_0^3\left(\frac{\varepsilon}{\pi Cd}\right)^{1/2}} - \sqrt{\frac{Cd}{\varepsilon\pi^3}} \right) - \frac{2\gamma}{R_0} \right. \tag{1.8}$$

The physical pressure acting on the sensor, however, is equal to P plus the pressure that is needed to deform the metal diaphragm. That additional pressure was determined experimentally. Figure 1.6 below shows the relationship between the pressure (in kPa) and the displacement in millimeters for the diaphragm that is used in the prototype described here.[1]

As can be seen from Figure 1.6, these metal diaphragms have excellent linearity in the region of low pressure. The operation of the prototype described here is actually entirely within that linear region. (Note, however, that this is only a part of the overall pressure acting on the sensor, and, as can be seen from Eq. (1.8), the sensor is actually nonlinear. More specifically, the pressure range for the diaphragm is only 0.6 kPa, while the pressure range for the entire sensor, taking into account the pressure required for deforming the mercury drop, is 3 kPa.) The slope of the linear relationship in Figure 1.6 can been determined from the data shown. Furthermore, since the capacitance of the device is directly proportional to the displacement of

---

[1]The diaphragm was mounted on a special fixture with a transducer that measures displacement. The fixture was then tested inside a pressure chamber, as discussed in the experimental results section.

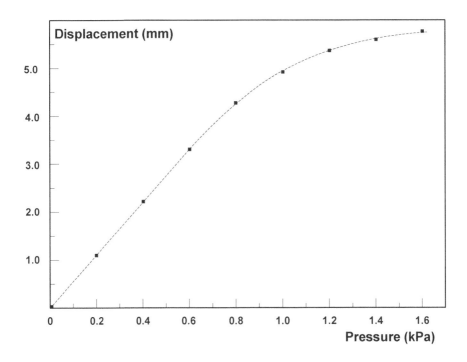

Figure 1.6: Pressure–displacement relationship for the metal diaphragm used in the prototype sensor.

the diaphragm[2] (specifically, a displacement of 3 mm corresponds to a capacitance change from 20 nF to 6.76 μF), the relationship between the applied pressure and the capacitance can be easily determined as well. That relationship can be represented as

$$P_{dia} = \alpha C \tag{1.9}$$

where P_{dia} is the pressure acting on the diaphragm and α is the constant of proportionality. For the prototype device described here, α was found to be equal to 0.09 kPa/μF. The physical, or total, pressure acting on the sensor is equal to the sum of the two pressures in Eqs. (1.8) and (1.9). That total pressure is now finally given by

$$P_{total} = \alpha C - \frac{2\gamma}{R_0} + 2\gamma / \left(\sqrt{\frac{Cd}{\varepsilon\pi^3} + \frac{4}{3}R_0^3 \left(\frac{\varepsilon}{\pi Cd}\right)^{1/2}} - \sqrt{\frac{Cd}{\varepsilon\pi^3}} \right) \tag{1.10}$$

By knowledge of the physical parameters of the device and by measuring the capacitance between the two external electrodes, the total pressure can therefore be

---

[2]Note from Eq. (1.5) that C is proportional to the contact area A only, since both ε and the dielectric thickness d are constants. A, in turn, is linearly related to the displacement of the diaphragm, since the volume of the mercury drop remains constant.

directly calculated. For the prototype sensor, the diaphragm pressure P_{dia} was found to reach a maximum of 0.6 kPa (see Figure 1.6), while the pressure P required to fully deform the mercury droplet was found to be about 2.4 kPa. The sensor can therefore handle a total pressure of about 3 kPa. Of course, for applications where a larger pressure range is needed, a stiffer diaphragm must be used.

1.1.3 Experimental Results

For the prototype sensor described here, the following are the dimensions and the physical constants:

- Dielectric constant (ε_r) of the barium strontium titanate ceramic layer: 12,000

- Thickness of the ceramic layer: 1 μm

- Radius R_0 of the undeformed mercury drop: 1.5 mm

- Wetting area of the undeformed mercury drop: 0.196 mm^2

- Wetting area of the fully deformed mercury drop: 63.6 mm^2

Notice that the ratio between the wetting areas in the deformed and the undeformed configurations is more than 300! In fact, straightforward substitution into Eq. (1.5) gives capacitances of 20 nF and 6.76 μF in the two cases, and direct experimental measurements have confirmed these numbers.

Figure 1.7 below shows a plot of the total pressure calculated from Eq. (1.10) as a function of the capacitance C, along with the actual pressure that was measured in a pressure chamber (several different samples were tested, and the readings of all samples were essentially the same).

The measurements were performed in a commercial-quality pressure chamber[3] that is pressurized with compressed air (room air). The chamber is equipped with two different types of commercially available, precalibrated pressure sensors[4] to eliminate any possibility of errors in the measurements. The capacitance was measured directly with a capacitance meter that was connected to the sensor to eliminate any possibility of errors in the estimation of capacitance. As the graphs in Figure 1.7 show, the agreement between the experimental and the theoretically calculated values of the pressure was excellent in the region of low pressures. However, a discrepancy of approximately 3% was observed in the region of high pressure (2–3 kPa). Since the discrepancy is not the result of measurement errors, the basic theoretical model of Eq. (1.10) will have to be refined. More specifically, it should be noted that the relationship between the displacement and the pressure acting on the diaphragm in Figure 1.6 is in fact not perfectly linear in the operating region of 0–0.6 kPa, and hence a polynomial representation instead of the simple constant α in Eq. (1.9) must

---

[3]Chamber type 175-10000, from Allied High Tech products, Inc.

[4]61CP Series Ceramic Capacitive Pressure Sensor, from Sensata Technologies, Inc; and MAX1450 evaluation kit, with a GE NovaSensor pressure sensor, from Maxim Integrated Circuits, Inc.

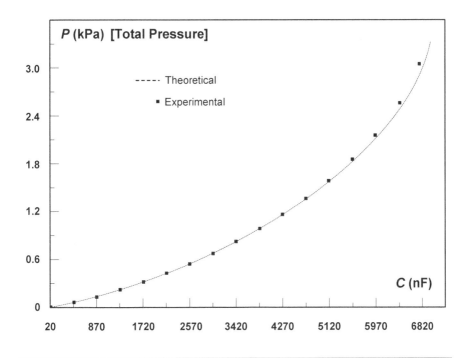

Figure 1.7: Total pressure acting on the sensor as a function of the measured capacitance.

be used. This refinement will not be given here, however, as such a polynomial relationship is strongly a function of the diaphragm used in the sensor (such a polynomial will vary depending on the type of metal, size, and structure of the diaphragm used). Alternatively, a calibration curve for the sensor can simply be used.

Temperature related errors:

For large changes in temperature, the error in the calculated pressure for the present sensor is actually slightly higher than other known types of capacitive pressure sensors due to the thermal expansion of the drop of mercury. The linear expansion of any metal object is given by [21]

$$L = L_0 (1 + \lambda \Delta T) \qquad (1.11)$$

where L is any linear dimension, λ is the metal's expansion coefficient, and ΔT is the change in temperature. The expansion in volume will be therefore given by

$$V = L^3 = L_0^3 (1 + 3\lambda \Delta T + 3\lambda^2 \Delta T^2 + \lambda^3 \Delta T^3) \qquad (1.12)$$

The coefficient of thermal expansion for mercury has a value of $9.1 \times 10^{-5} °C^{-1}$. Therefore terms containing λ^2 and higher powers can be easily neglected. The above equation then takes the more simplified form

$$V \approx V_0 (1 + 3\lambda \Delta T) \qquad (1.13)$$

To relate Eq. (1.13) to the expansion of the radius R of the mercury droplet, the equation can be written as follows:

$$\frac{\Delta V}{V_0} = \frac{(\Delta R)^3}{R_0^3} = 3\lambda \Delta T \tag{1.14}$$

The volume of the deformed droplet is approximately equal to area $\times$ thickness (see Figure 1.5). Hence, for any specific deformation (where the thickness is held constant), the change in the contact area ΔA can be represented as follows:

$$\frac{\Delta A}{A} = \frac{(\Delta R)^3}{R_0^3} = 3\lambda \Delta T \tag{1.15}$$

Furthermore, the ratio $\Delta A/A$ is equal to $\Delta C/C$, and from Figure 1.7 it is clear that very small variations in capacitance can be approximated as a linear function of the pressure variation; hence

$$\frac{\Delta A}{A} = \frac{\Delta C}{C} \approx \frac{\Delta P}{P} = 3\lambda \Delta T \tag{1.16}$$

Straightforward substitution into the equation at $T = 80°C$ ($\Delta T = 55°C$, in comparison with room temperature) gives a ratio $\Delta P/P$ of 0.015, or 1.5%. Another substitution for $T = 40°C$ ($\Delta T = 15°C$) gives $\Delta P/P = 0.41\%$.

Figure 1.8 shows the error ΔP that was measured (as a percentage) as the temperature varied in the range of $+10°C$ to $+40°C$ for each of the values of the pressure shown in Figure 1.7. As shown, the maximum deviation observed was indeed 0.4%, in very good agreement with the theoretical prediction. A similar set of measurements for the temperature range of $-10°C$ to $+80°C$ confirmed the theoretical prediction of a maximum deviation $\Delta P/P$ of 1.5%. Figure 1.9 shows temperature hysteresis curves that were obtained at a fixed pressure of 3 kPa (a, b), and at a fixed pressure of 0.1 kPa (c, d). The figures show the pressure that was measured as the temperature was cycled, and essentially confirm the same results.

It is very important to point out that it is possible to totally compensate for the error in the measured pressure by using Eq. (1.16) if the temperature is known. Finally, it is important to point out that the effects discussed above are totally due to the thermal expansion of the drop of mercury. The stainless steel diaphragm used in the sensor has a very low expansion coefficient and its contribution to the error in the value of P was found to be negligible.

Pressure hysteresis:

The hysteresis in the values of the calculated pressure was determined by cycling the pressure at a fixed temperature and plotting the measured pressure versus the actual applied pressure. The results are shown in Figure 1.10 for two different extreme values of temperature. As the plots in Figure 1.10 show, The maximum hysteresis was found to be $\pm 0.05\%$, essentially negligible. (It is to be pointed out that the sensor was subjected to 1000 pressure cycles in each test, and the results did not vary significantly).

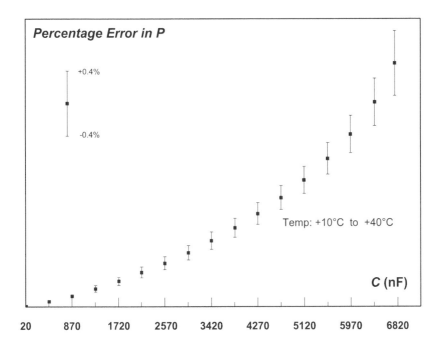

Figure 1.8: The ratio $\Delta P/P$ (expressed as a percentage, with an arbitrary scale) for each of the values of the nominal pressure shown in Figure 1.7. The error is the result of a maximum deviation of $\pm 15°$C from room temperature.

Susceptibility to mechanical shocks:

If the sensor is shocked, the mercury drop will be momentarily displaced (especially if a small pressure is acting on the drop), and it was observed that a "recovery time" is needed for the drop to return to its original position (and hence for the momentary error to disappear). The sensor was tested on a standard electrodynamic shaker[5] that provides shock pulses of a magnitude of 34 g and a duration of 10 ms [22]. To determine the recovery time that is needed for the error to disappear after the shock, the frequency of the built-in 555 oscillator was monitored with a digital storage oscilloscope. The nominal frequency (the frequency before the shock) was noted, and the time duration that lapsed from the onset of the shock until the frequency returned to its nominal value was observed on the scope. The results of that test are shown in Figure 1.11. As the figure shows, the recovery time was negligible for large pressures. This indicates that the mercury drop does not suffer from significant displacements when it is pressurized. When the pressure acting on the sensor is small, however, the recovery time is much more significant, as the figure shows. The longest recovery time observed was about 50 ms. It should be finally mentioned that while the sensor was mounted on the electrodynamic shaker, it was pressurized

---

[5]Model DSS-M100, from Dynamic Solutions, Inc.

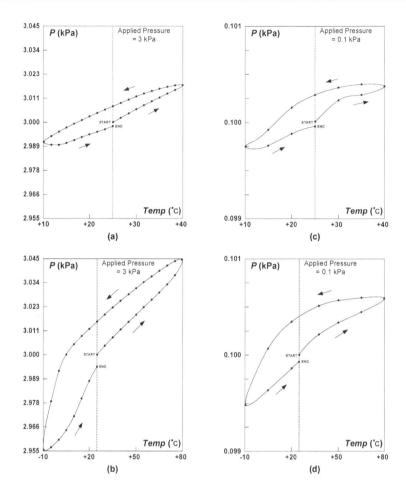

Figure 1.9: Measured pressure vs. temperature for a fixed applied pressure of 3 kPa (a and b) and for a fixed applied pressure of 0.1 kPa (c and d). Each figure shows one complete cycle, starting at 25°C.

with a small mechanical fixture that uses a screw for exerting force on the steel diaphragm. By measuring the capacitance of the sensor, the pressure in each phase of the test was determined precisely.

Environmental effects:

At elevated temperatures (a few hundred°C), mercury reacts with the oxygen in the air to form mercury oxide. The presence of mercury oxide can severely degrade the performance of the sensor. For this reason, the sensor is not intended to function at elevated temperatures. (Another issue to consider is the expansion of the mercury drop, as indicated above). Generally, the use of this sensor at temperatures above +80°C is not recommended.

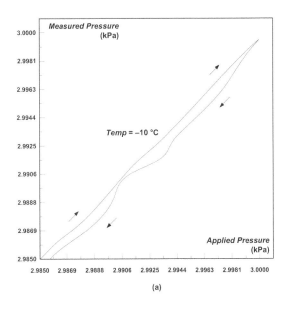

(a)

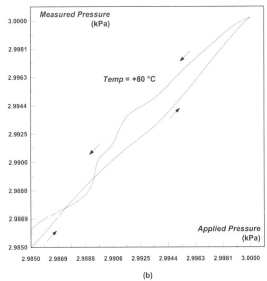

(b)

Figure 1.10: Pressure hysteresis curves at −10°C and +80°C in the vicinity of full-scale pressure (3 kPa). The maximum hysteresis error is less than ± 0.05% FSO.

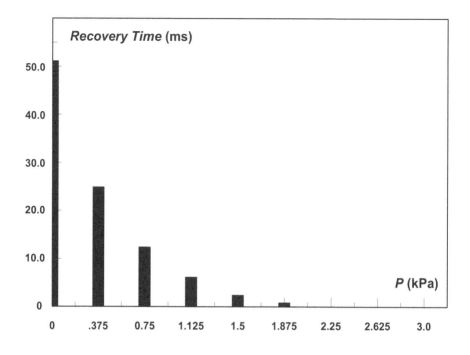

Figure 1.11: After-shock recovery time as a function of the applied pressure.

1.2 Inductive Pressure Sensors

Pressure sensors exist in mainly two varieties: piezoresistive and capacitive. Piezoresistive pressure sensors are characterized by good linearity and acceptable sensitivity, but the temperature hysteresis in these sensors is usually quite large [3, 10]. Capacitive pressure sensors can achieve much higher sensitivity and lower temperature hysteresis, as pointed out earlier, but they are usually nonlinear [1, 9]. A new type of pressure sensor was introduced recently in the literature. The sensor is characterized by (1) miniature size (the sensor fits inside an IC package); (2) excellent linearity over an arbitrarily chosen pressure range; (3) substantially high sensitivity; and (4) substantially low temperature hysteresis. This new type of sensor is also currently in production and should be commercially available in early 2015.

1.2.1 Structure

Unlike piezoresistive and capacitive pressure sensors, the new sensor is based on a technique for substantially changing the inductance of a coil. The principle behind the new sensor is shown in Figure 1.12.

The principle of the new device, as shown in Figure 1.12, is to create a highly variable inductance mechanism. In the mechanism shown in Figure 1.12, the change in inductance is equal to the relative magnetic permeability of the core material and is

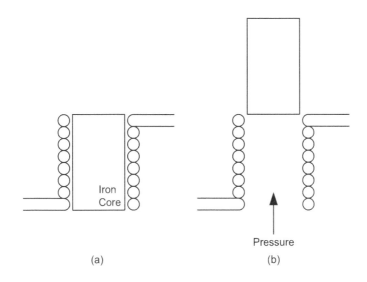

Figure 1.12: (a) A movable iron core is positioned inside the core of a vertical inductor (coil). (b) As pressure acts on the iron core in the vertical direction, the core can
be totally displaced outside the coil. The change in inductance between the two configurations is equal to the relative magnetic permeability of the core material and is
typically 4000-fold or higher.

typically 4000-fold or higher. By comparison with the LVDT, this sensor offers substantially higher sensitivity, does not require AC excitation, and since it is composed
of only one coil, it is substantially more compact than a typical LVDT [27–29]. The
prototype device that was built and tested by the author has demonstrated a change
in inductance of approximately 34.5 mH over a pressure range of 0.3 to 10 kPa. Its
sensitivity is therefore 3.56 mH / kPa, which is substantially high.

Table 1.2 lists the four most important parameters of the new sensor: sensitivity,
linearity, pressure hysteresis, and temperature hysteresis, as compared to the other
known types of pressure sensors. In addition to the parameters shown in the table,
other important aspects of comparison include cost and size. Piezo-resistive and capacitive pressure sensors are generally miniature sensors, with cost that ranges from
a fraction of a dollar up to a few dollars. LVDT-based pressure sensors, on the other
hand, are large, bulky sensors [27–29], with a cost that generally exceeds $100. The
new sensor introduced here is of the miniature, low-cost variety.

The structure of the new inductive pressure sensor is shown schematically in
Figure 1.13. Figure 1.14 is a photograph of the actual components of the sensor.

As shown in Figures 1.13 and 1.14, a vertical coil of a height of 4 mm and a diameter of 12 mm is totally embedded inside an open-cavity, 24-pin DIP IC package.
A small cylindrical iron core of a height of 4 mm and a diameter of 6 mm is positioned inside the coil, surrounded by a smooth Teflon sleeve. In the device shown,
the pressure acts on the iron core in the upward direction. However, this is only one
possible configuration, and it is equally possible to configure the sensor such that the

Table 1.2 Comparison of the new sensor to the other known types of pressure sensors (see [10,11]). Note that uncertainty in the sensitivity data is typically not provided by manufacturers.

	Sensitivity	Linearity	Pressure Hysteresis	Temperature Hysteresis ($-10°C$ to $+80°C$)
Piezoresistive pressure sensors	up to 25 mV/kPa	Generally linear	up to $\pm$ 1% FSO	up to $\pm$ 2% FSO
Capacitive pressure sensors	up to 0.2 nF/kPa	Generally nonlinear	up to $\pm$ 0.1% FSO	up to $\pm$ 0.5% FSO
LVDT pressure sensors	up to 400 mV/kPa	Generally linear	up to $\pm$ 0.5% FSO	up to $\pm$ 0.1% FSO
New sensor (uncompensated)	3.56 mH/kPa or higher	Linear	less than $\pm$ 0.05% FSO	up to $\pm$ 0.1% FSO

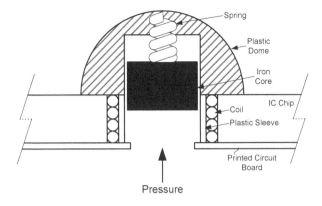

Figure 1.13: Mechanical structure of the sensor (scale: 3:1).

Figure 1.14: Components of the sensor. The coil is totally mounted inside a standard 24-pin DIP IC package (dimensions: 30 mm × 14 mm). The iron core and the plastic sleeve are positioned inside the core of the coil, and the plastic dome shown in the figure is mounted on top of the coil and attached permanently with an adhesive.

pressure acts downward. Also as shown in the figures, a semi-spherical plastic dome is positioned on top of the coil in order to contain the iron core it displaced. In the internal cavity of the dome (see Figure 1.13), a spring with a known spring constant is mounted. As the displaced iron core exerts force on the spring, the displacement will be proportional to the force (and hence pressure) that is acting on the iron core.

The displacement of the iron core, in addition, will be related to the observed inductance of the coil. These two characteristics allow the pressure acting on the sensor to be calculated as a function of the observed inductance.

1.2.2 Theory

Inductance as a function of the position of the iron core:

The inductance L of any inductor is given by the well-known equation [24]

$$L = \frac{\mu_0 \mu_r N^2 A}{l} \tag{1.17}$$

where μ_0 is the magnetic permeability of free space, μ_r is the relative permeability of the material present in the core, N is the number of turns in the coil, A is the cross-sectional area of the coil, and l is its length. Figure 1.15 shows the creation of two inductors in series as the iron core is displaced by a small distance x from its original position. The first inductor contains the iron core and its inductance is labeled L_1, and the second inductor contains only air in its core and its inductance is labeled L_2. As the figure shows, the total length of the coil in the device is l.

Based on the above formula, the inductances L_1 and L_2 will be now given in terms of the displacement x as follows:

$$L_1 = \frac{\mu_0 \mu_r N_1^2 A}{(l - x)}$$

$$L_2 = \frac{\mu_0 N_2^2 A}{x} \tag{1.18}$$

where N_1 is the number of turns in the first inductor and N_2 is the number of turns in the second inductor. The following two relationships now must hold:

$$N_1 + N_2 = N \quad \text{and} \tag{1.19}$$

$$\frac{N_2}{N} = \frac{x}{l} \tag{1.20}$$

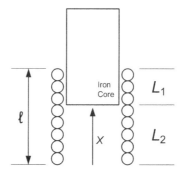

Figure 1.15: As a result of the displacement of the iron core, the coil is effectively split into two inductors connected in series.

These two relationships can be alternatively written as follows:

$$N_1 = N\left(1 - \frac{x}{l}\right)$$

$$N_2 = N\left(\frac{x}{l}\right) \tag{1.21}$$

By substitution for N_1 and N_2 from the above two identities into the equations in (1.18) we obtain, after simplification of the results,

$$L_1 = \frac{\mu_0 \mu_r N^2 A}{l}\left(1 - \frac{x}{l}\right)$$

$$L_2 = \frac{\mu_0 N^2 A x}{l^2} \tag{1.22}$$

From circuit theory, the total inductance of a series combination of two inductors is equal to the sum of the individual inductances. By taking the sum $L_1 + L_2$ and simplifying the resulting expression, we obtain,

$$\begin{aligned} L = L_1 + L_2 &= \frac{\mu_0 N^2 A}{l}\left[\mu_r - \frac{x}{l}(\mu_r - 1)\right] \\ &= L_0\left[\mu_r - \frac{x}{l}(\mu_r - 1)\right] \end{aligned} \tag{1.23}$$

where L is the total observed inductance and L_0 is the inductance of the coil with an air core (minimum inductance). By using the parameters of the present device (see Section 1.4), Figure 1.16 shows a plot of L as a function of the displacement x.

As expected, the observed inductance varies linearly from a maximum value L_{max} (inductance of the coil with iron core) to a minimum value L_0 (which has a value of 11.5 μH in the present prototype, and therefore appears negligible by comparison with L_{max}). As can be seen from Eq. (1.23), the ratio $L_{max}/L_0 = \mu_r$, which is approximately equal to 4000 for iron. The plot in Figure 1.21 also shows the experimentally measured values of L. As can be seen, the plot actually deviates from the theoretically expected linear behavior when L is very close to L_0. This is due to the simplifying assumptions of the theory of inductance (specifically, the core of the coil is not actually purely air when the iron cylinder is placed in the immediate vicinity of the coil). Nevertheless, the linear response is obtained for most of the dynamic range of the sensor. In the present prototype, the region of operation is only the linear region (where $0\,\text{mm} \leq x \leq 3\,\text{mm}$).

Position of the iron core as a function of the applied pressure:

In the present prototype, the pressure acts on the iron core in the upward direction. Accordingly, a minimum force $F_{min} = mg$ must be applied, where m is the mass of the iron core and g is the acceleration of gravity. The minimum pressure P_{min} that must be applied to the sensor is therefore equal to

$$P_{min} = \frac{F_{min}}{A} \tag{1.24}$$

The sensor cannot respond to any pressure less than P_{min}, which is approximately

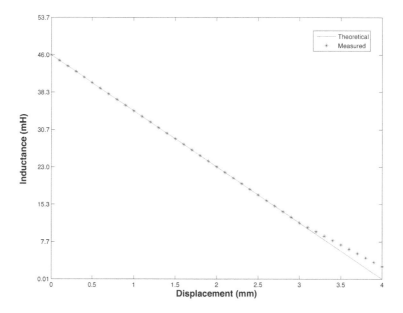

Figure 1.16: Total inductance (in mH) as a function of the displacement x, according to Eq. (1.23).

equal to 0.3 kPa in the present prototype. Pressures larger than P_{min} will result in a force that will compress the vertical spring (see Figure 1.13). In this case, the net force acting to compress the spring will be given by

$$F = kx = (P - P_{min})A \tag{1.25}$$

where k is the spring constant and P is the pressure acting on the sensor. Hence, the displacement x of the iron core will be given as a function of pressure by

$$x = \frac{(P - P_{min})A}{k} \tag{1.26}$$

Applied pressure as a function of the observed inductance L:

By substituting from Eq. (1.26) into Eq. (1.23), we obtain a relationship between L and the applied pressure:

$$L = L_0 \left[\mu_r - (\mu_r - 1)\frac{(P - P_{min})A}{kl} \right] \tag{1.27}$$

Solving for P, we obtain

$$P = P_{min} + \frac{kl}{A}\left(\frac{\mu_r - L/L_0}{\mu_r - 1}\right) \tag{1.28}$$

It should be pointed out that the sensor is essentially a differential pressure sensor (since ambient air exists on both sides of the moving iron core). The pressure calculated from Eq. (1.28) is thus the pressure that exceeds atmospheric pressure. To obtain a measurement of absolute pressure, the atmospheric pressure must simply be added to the pressure calculated from Eq. (1.28).

1.2.3   Experimental Results

Basic results:

For the prototype device described here, the following are the dimensions and the physical constants:

- Spring constant $k = 91.33$ N/m

- Relative permeability of the iron core (electrical steel) $\mu_r = 3978$

- Diameter of the core/iron core = 6 mm

- Height of the iron core and height of the coil $l = 4$ mm

- Mass of the iron core $m = 0.89$ g

- Total number of turns in the coil $N = 36$

Based on the above data, the minimum inductance $L_0 = 11.5$ μH and the maximum inductance L_{max} is found to be equal to 46 mH (Eq. (1.17)). From Eq. (1.24), $P_{min} = 0.3$ kPa, and from Eq. (1.26) P_{max} is found to be approximately equal to 10 kPa (at a displacement $x = 3$ mm). The range of 0.3 kPa to 10 kPa is therefore the dynamic range in the present prototype (of course, the dynamic range can be changed by varying the parameters, particularly the spring constant). Figure 1.17 shows a plot of the pressure P predicted by Eq. (1.28) as a function of the measured inductance L, along with the actual pressure values that were measured in a pressure chamber.

The measurements were performed in a commercial-quality pressure chamber[6] that is pressurized with compressed air. The chamber is equipped with two different types of commercially available, pre-calibrated pressure sensors[7] to eliminate any possibility of errors in the measurements. The inductance was measured directly with an LRC meter that was connected to the sensor. As Figure 1.17 shows, the agreement between the experimental and the theoretically calculated values of the pressure is excellent (it is to be pointed out that the lowest inductance shown in the plot is approximately 11.5 mH to avoid the nonlinear region near L_0; see Figure 1.16). Within the linear region shown in the figure, it can be clearly seen that the dynamic range of the present sensor is indeed 0.3 kPa to 10 kPa. It can also be seen that the sensitivity (or change in inductance per unit pressure) is about 3.56 mH/kPa.

---

[6]Chamber type 175-10000, from Allied High Tech products, Inc.
[7]61CP Series Ceramic Capacitive Pressure Sensor, from Sensata Technologies, Inc; and MAX1450 evaluation kit, with a GE NovaSensor pressure sensor, from Maxim Integrated Circuits, Inc.

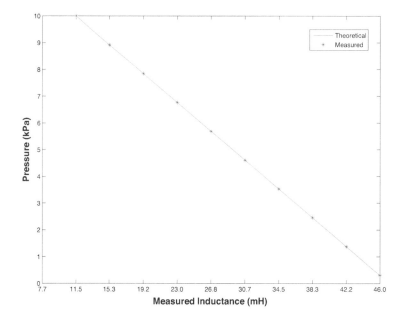

Figure 1.17: Pressure *P* as a function of the measured inductance *L*.

Pressure hysteresis:

The hysteresis in the values of the calculated pressure was determined by cycling the pressure applied to the sensor at a fixed temperature and plotting the measured pressure versus the actual applied pressure. The results are shown in Figure 1.18 for two different extreme values of temperature. As the plots in Figure 1.18 show, the maximum hysteresis error was found to be ±0.05%, essentially negligible. (It is to be pointed out that the sensor was subjected to 100 pressure cycles in each test, and the results did not vary significantly.) The pressure hysteresis in this sensor is almost totally due to the "memory effect" in the spring.

Temperature hysteresis:

Figure 1.19 shows temperature hysteresis curves that were obtained at a fixed pressure of 10 kPa (a, b) and at a fixed pressure of 0.3 kPa (c, d). As the figures clearly show, the maximum temperature hysteresis error is about ± 0.1% FSO. The temperature hysteresis is believed to be totally due to variations in the relative permeability of the iron core as the temperature is cycled.

Susceptibility to mechanical shocks:

If the sensor is shocked in the vertical direction, the iron core will be momentarily displaced, and it was observed that a "recovery time" is needed for the iron core to return to its original position (and hence for the momentary error to disappear). The

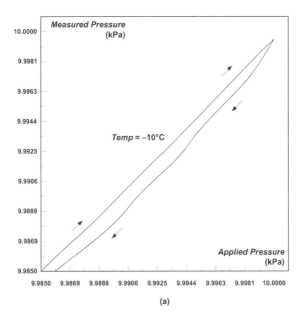

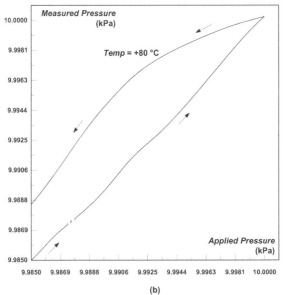

Figure 1.18: Pressure hysteresis curves at −10°C and +80°C, in the vicinity of the full-scale pressure (10 kPa). The maximum hysteresis error is less than ± 0.05% FSO.

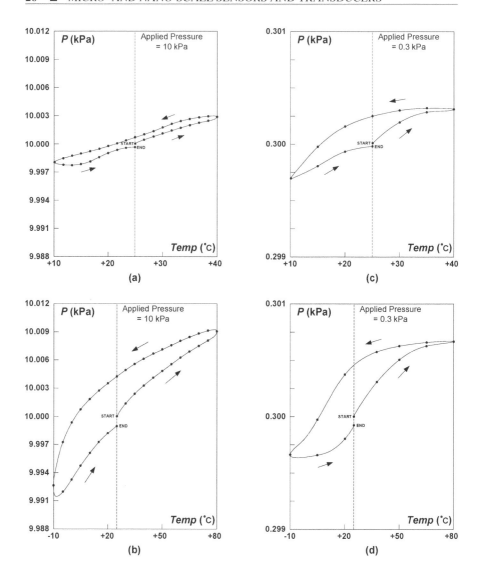

Figure 1.19: Measured pressure vs. temperature for a fixed applied pressure of 10 kPa (a and b) and for a fixed applied pressure of 0.3 kPa (c and d). Each figure shows one complete cycle, starting at 25°C.

sensor was tested on a standard electrodynamic shaker[8] that provides shock pulses of a magnitude of 34 g and a duration of 10 ms [21, 22]. In order to determine the recovery time that is needed for the error to disappear after the shock, the frequency of an interface circuit that uses a 555 oscillator (see Section 1.5 below) was monitored

---

[8]Model DSS-M100, from Dynamic Solutions, Inc.

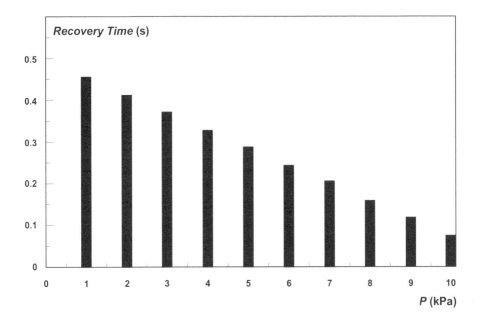

Figure 1.20: After-shock recovery time as a function of the applied pressure.

with a digital storage oscilloscope. The nominal frequency of the interface circuit was noted, and the time duration that lapsed from the onset of the shock until the frequency returned to its nominal value was observed on the scope. The results of that test are shown in Figure 1.20. It should be mentioned that while the sensor was mounted on the electrodynamic shaker, it was pressurized with a small mechanical fixture that uses a screw for exerting force on the iron core. By measuring the inductance of the coil, the pressure in each phase of the test was determined precisely.

1.2.4 Sensor Interface Circuit

A word is now in order concerning the interface circuit that is used with the sensor. An interface circuit that uses a 555 timer chip working in an oscillator mode was integrated onto the printed circuit board on which the sensor is mounted. The oscillator circuit essentially converts the unknown inductance of the coil to frequency. Such a circuit is very well known in the literature and is described in references such as [18]. The block diagram of the circuit is shown in Figure 1.21.

The equation that characterizes the circuit shown in Figure 1.21 is [18]

$$f = 0.092\frac{R}{L} \qquad (1.29)$$

where f is the frequency of the resulting square wave in Hz. Given a known resistance R, the value of the unknown inductance L can be easily calculated from this

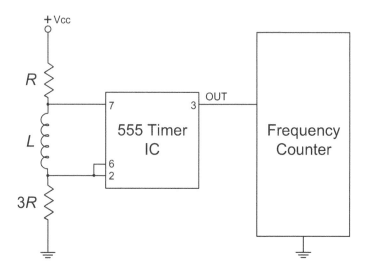

Figure 1.21: Block diagram of the interface circuit used to measure the inductance *L*.

equation by simply observing the frequency. This is how plots such as the plots in Figures 1.16 and 1.17 were obtained. It is to be pointed out that the interface circuit does not amplify or compensate for the characteristics of the sensor (i.e., the sensor is uncompensated, as shown in Table 1.2).

1.3 Ultrahigh Sensitivity Pressure Sensors

Recent trends in earthquake monitoring and prediction have created a requirement for a highly sensitive vibration detector (e.g., a detector with sensitivity for accelerations of less than 10^{-6} g) [30–35]. In addition, a highly sensitive pressure sensor (a sensor with a capability to detect pressures of less than one Pascal) will be very useful for a new class of biological and molecular sensing applications [36, 37]. As indicated earlier, sensors based on variable capacitance can offer the highest sensitivity among all known types of pressure/vibration sensing mechanisms [7,8,10,11,38]. However, capacitive pressure/vibration sensors still cannot offer the ultrahigh sensitivity that is needed for demanding applications such as earthquake prediction and the detection of very small pressures in biological processes. This section introduces such a new sensor with ultrahigh sensitivity. Like the earlier two types of pressure sensors, this type of sensor is also currently in production and should be commercially available in early 2015.

1.3.1 Structure

The basic idea behind the new sensor is to create a transduction mechanism that uses a variable ultracapacitor rather than a variable capacitor. In this mechanism, an extremely small displacement of 20 μm (less than the width of a human hair) triggers a substantially large variation in capacitance. This concept is shown in Figure 1.22.

Hence, according to the concept in Figure 1.22, an extremely small displacement of 20 μm results in a substantially large variation in capacitance. This is due to the capacitance of the two-electrode arrangement being equal to zero when the movable carbon nanotube (CNT) electrode is outside of the electrolyte, while full capacitance is obtained as the CNTs are fully immersed in the electrolyte. The sensitivity of this transduction mechanism is therefore unmatched among all the known types of pressure/vibration sensors. It is to be pointed out that carbon is hydrophobic and therefore droplets of the electrolyte do not attach to the CNTs in this application. This results in a smooth linear variation in capacitance as the movable electrode travels the very short distance of 20 μm (see the experimental results).

Figure 1.23 is a mechanical diagram showing the actual construction of the sensor. The rubber membrane to which the movable electrode is attached acts as a spring mechanism that permits the movable electrode to retreat back to its original position.

Figure 1.24 is a photograph of the actual sensor. The rubber membrane used is actually conductive rubber, and therefore there is no need to connect a wire terminal directly to the movable electrode. The fixed electrode (not shown in the picture) is mounted underneath the movable electrode and is connected to its own terminal. The entire sensor is housed inside an off-the-shelf stainless steel enclosure that provides two external terminals for easy connection to other circuitry. These are the terminals of the variable capacitor. One stand-alone electrode is shown in the figure next to the sensor (a layer of CNTs is grown on one side of the electrode. These CNTs were deposited by a specialized commercial supplier that utilizes the process of chemical vapor deposition [39, 40]). Figure 1.25 is a scanning electron microscope (SEM) micrograph that shows the CNTs deposited on the surface of the electrode.

1.3.2 Theory

Sensing of pressure:

As a slight pressure is applied on the upper electrode, the length of the CNTs that is immersed in the electrolyte increases, and hence the capacitance between the two electrodes increases. We shall now obtain an expression that relates the capacitance to the displacement x, and accordingly to the pressure applied to the sensor. The surface area A of the CNTs that is immersed in the electrolyte will be given by

$$A = N \times 2\pi r x \qquad (1.30)$$

where N is the total number of CNTs on the surface of the electrode, r is the radius of one carbon nanotube, and x is the displacement of the electrode (the frontal area of the carbon nanotube is negligible in comparison with the lateral area and will be ignored in this analysis). The surface density of CNTs in the present application is 10^9 tubes per cm^2, and the area of the electrode is 0.78 cm^2. The total number N of

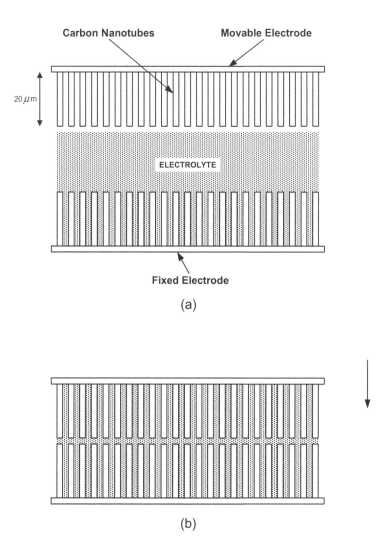

Figure 1.22: Fundamental concept of the new ultrahigh sensitivity pressure/vibration sensor. (a) An ultracapacitor consisting of two electrodes is assembled such that one electrode is fixed and is fully immersed in the electrolyte while the other electrode is movable and is initially positioned outside of the electrolytic solution. Each electrode consists of a stainless steel plate on which carbon nanotubes of a length of approximately 20 μm are grown (the length of the carbon nanotubes is greatly exaggerated in the drawing). (b) As pressure or vibration is applied, the movable electrode travels downward and dips into the electrolytic solution. As the electrode travels a distance of 20 μm (the length of the carbon nanotubes), the capacitance increases from zero to full capacitance (approximately 54 μF in the present prototype).

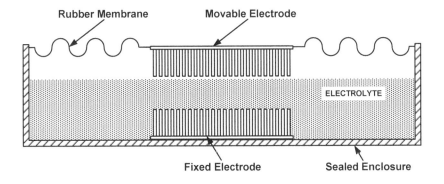

Figure 1.23: Mechanical structure of the sensor (not to scale).

Figure 1.24: Photograph of the sensor. The integrated package measures 23 mm (width) × 30 mm (height).

CNTs on the surface of the electrode is therefore approximately 0.78×10^9 CNTs. The diameter of one CNT in the present application is about 250 nm.[9] From this data, the immersed area $A = 612.6x$, where x is in meters. The overall capacitance C of the ultracapacitor is a series combination of the capacitances at each of the electrode–electrolyte interfaces [41, 42], i.e.,

$$C = \frac{1}{2} \frac{\varepsilon_0 \varepsilon_r A}{d} \tag{1.31}$$

where ε_0 is the permittivity of free space, ε_r is the relative permittivity of the electrolyte,[10] and d is the electrode–electrolyte separation.[11] The above expression is the

---

[9]The diameter of the CNTs in the present application is larger than in most other applications, since sturdy CNTs are essential for the long term reliability of the sensor.

[10]For ultracapacitors, the relative permittivity of the 1 nm thick electrolyte layer at the interface is approximately equal to 1.

[11]This separation is about 1 nm in ultracapacitors [41, 43]. This separation essentially results due to the hydrophobic nature of the CNTs, which prevents the charged ions in the electrolyte from reaching the carbon surface.

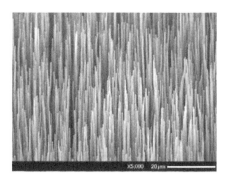

Figure 1.25: Multiwalled carbon nanotubes of an average diameter of 250 nm and a length of about 20 μm, grown on a stainless steel electrode.

well-known equation of a series combination of two parallel-plate capacitors. It is important to note here that the capacitance of a cylindrical capacitor is very nearly the same as the capacitance of a parallel-plate capacitor when the separation between the positive and the negative charges is on the order of a few nanometers. By taking $d = 1$ nm, $\varepsilon_r \approx 1$, and substituting with $A = 612.6x$ in the above equation, we get

$$C = 2.71x \tag{1.32}$$

For a maximum displacement of 20 μm, the capacitance C according to the above equation will be approximately 54 μF. The force F acting on the sensor is equal to kx, where k is the spring constant of the rubber membrane [44]. In the present sensor, $k \approx 3$ N/m; hence,

$$F = 3\left(\frac{C}{2.71}\right) = 1.11\,C \qquad \text{newton} \tag{1.33}$$

Since a capacitance of a few picofarads is easily measurable with electronic circuitry, it can be seen that the sensor is sensitive to forces on the order of a few pN (typical scale of force in molecular and biological processes [36, 37]). The pressure acting on the sensor is simply given by the ratio

$$P = \frac{F}{A_{electrode}} \tag{1.34}$$

where $A_{electrode}$ is the area of the movable electrode. Since this area is 0.78 cm^2, we find that

$$P = 14230.77\,C \qquad \text{Pa} \tag{1.35}$$

Since the maximum capacitance is 54 μF, we conclude that the measurable pressure is in the range of 0 to 0.77 Pa. Once again, it can be seen that a pressure of only 14 nanopascal will result in a capacitance of 1 pF (a measurable capacitance).

Sensing of vibrations:

When the sensor is subjected to vibration, the moving electrode can be modeled as a forced harmonic oscillator. The standard solution of the problem of the forced harmonic oscillator is [44, 45]

$$x(t) = \frac{F_{max} \sin \omega t}{m \sqrt{(\omega_0^2 - \omega^2)^2 + \beta^2 \omega^2}} \tag{1.36}$$

where m is the mass of the oscillator (the electrode), $\omega_0 = \sqrt{k/m}$ is the natural frequency of oscillation, β is the damping coefficient, and $F_{max} \sin \omega t$ is the sinusoidal force acting on the oscillator. This force is given at any time by the sum $kx \pm \beta dx/dt$. This force is also equal to $ma(t)$, where $a(t)$ is the instantaneous acceleration experienced by the electrode. We shall therefore replace F_{max} in Eq. (1.36) by ma, where a is the peak value of the acceleration. Equation (1.36) then yields

$$x_{max} = \frac{a}{\sqrt{(\omega_0^2 - \omega^2)^2 + \beta^2 \omega^2}} \tag{1.37}$$

From Eqs. (1.32) and (1.37), we conclude that

$$a = \frac{C_{max}}{2.71} \sqrt{(\omega_0^2 - \omega^2)^2 + \beta^2 \omega^2} \tag{1.38}$$

If the sensor is subjected to a frequency $\omega = \omega_0$, the above expression will reduce to

$$a = \frac{\beta \omega_0}{2.71} C_{max} \tag{1.39}$$

β was determined to be approximately equal to 0.8 for the present sensor (see experimental results below). For a capacitance C_{max} in the picofarad range, it can be seen that the detectable acceleration will be on the order of 10^{-12} m/s^2, or 10^{-13} g. The sensitivity of this device is therefore unmatched among all the known types of acceleration sensors [46–49].

1.3.3 Experimental Results

Pressure measurement:

Figure 1.26 shows a plot of the pressure calculated from Eq. (1.35) as a function of the capacitance C, along with the actual pressure that was measured in a pressure chamber. It is to be pointed out that the sensor is essentially a differential pressure sensor (only pressure in excess of the atmospheric pressure is detected).

To detect the very minute amounts of pressure shown in Figure 1.26, a technique was developed in which two identical pressure chambers were used.[12] The first chamber is equipped with a vacuum gauge, and the second chamber contains the pressure sensor. Very small amounts of air were then pumped from the first chamber

---

[12]Chamber type 175-10000 from Allied High Tech Products, Inc.

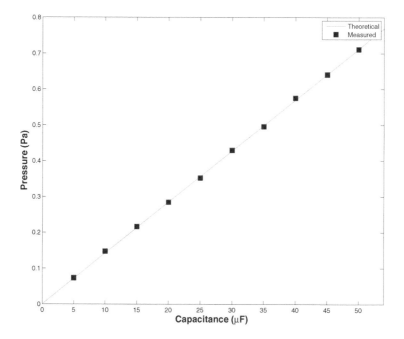

Figure 1.26: Pressure acting on the sensor as a function of the measured capacitance.

into the second. By reading the vacuum gauge's measurement in the first chamber, the pressure in the second chamber was calculated. The results are as shown in Figure 1.26.

Figure 1.27 shows temperature hysteresis curves that were obtained at a fixed pressure of 0.77 Pa (a, b), and at a fixed pressure of 14 nPa (c,d). The figures show the pressure that was measured as the temperature was cycled. The maximum hysteresis error is 1.5% FSO, as the figures show.

Figure 1.28 shows pressure hysteresis curves that were obtained by cycling the pressure at a fixed temperature and plotting the measured pressure versus the actual applied pressure. As the plots show, the maximum pressure hysteresis error is 0.4% FSO.

If the sensor is shocked, the moving electrode will be momentarily displaced, and it was observed that a "recovery time" is needed for the moving electrode to return to its original position (and hence for the momentary error to disappear). The sensor was tested on a standard electrodynamic shaker[13] that provides shock pulses of a magnitude of 34 g and a duration of 10 ms [9, 10]. In order to determine the recovery time that is needed for the error to disappear after the shock, the frequency of an interface circuit that uses a 555 oscillator was monitored with a digital storage oscilloscope. The nominal frequency of the interface circuit was noted, and the time duration that

---

[13] Model DSS-M100, from Dynamic Solutions, Inc.

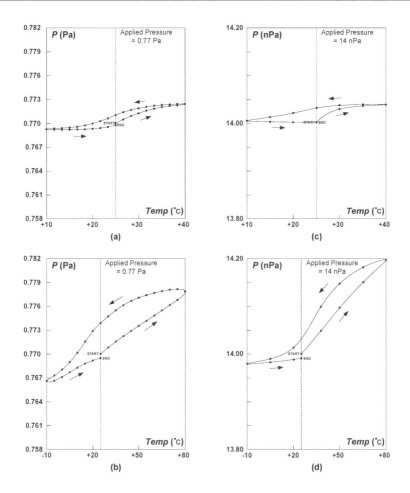

Figure 1.27: Measured pressure vs. temperature for a fixed applied pressure of 0.77 Pa (a and b) and for a fixed applied pressure of 14 nPa (c and d). Each figure shows one complete cycle, starting at 25°C.

lapsed from the onset of the shock until the frequency returned to its nominal value was observed on the scope. The results of that test are shown in Figure 1.29.

Figure 1.30 shows a block diagram of the interface circuit that was used in the various tests conducted. The circuit is a well-known 555 oscillator circuit that converts an unknown capacitance to frequency (the circuit is described in references such as [12]). It is to be pointed out that the interface circuit does not amplify or compensate for the characteristics of the sensor.

Acceleration measurement:

The sensor was subjected to vibrations at a frequency of 14 Hz on the electrodynamic shaker. This frequency is close to the natural frequency of the moving

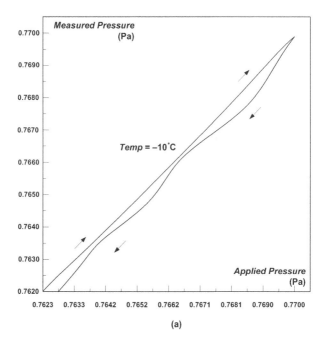

(a)

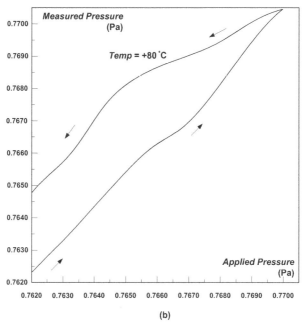

(b)

Figure 1.28: Pressure hysteresis curves at −10°C and +80°C in the vicinity of the full-scale pressure (0.77 Pa). The maximum hysteresis error is less than ± 0.4% FSO.

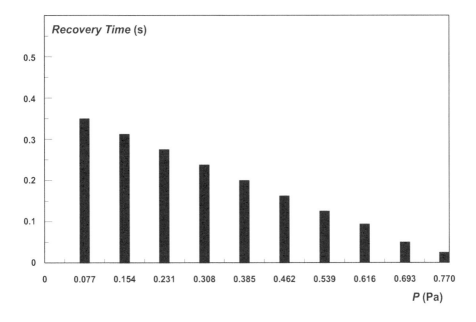

Figure 1.29: After-shock recovery time as a function of the applied pressure.

electrode (which was determined to be approximately 11.25 Hz for the present prototype). Figure 1.31 shows the measured capacitance as a function of time. It is to be observed that the moving electrode is practically outside of the electrolytic solution

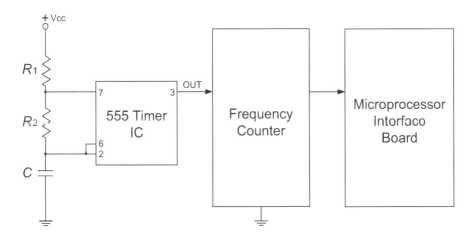

Figure 1.30: Block diagram of the interface circuit used to measure the capacitance C.

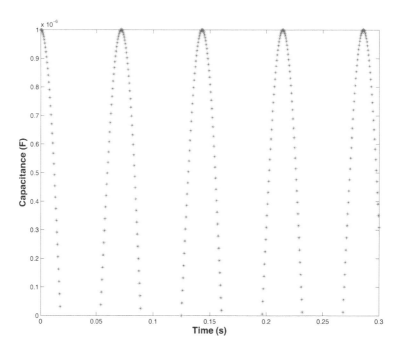

Figure 1.31: Measured capacitance as a function of time, as the sensor was subjected to vibrations at a frequency of 14 Hz.

for a half of the cycle, and, as shown, the capacitance is indeed zero for one-half of each cycle.

To test the sensor's response to acceleration, the sensor was mounted on a different electrodynamic shaker (from Data Physics Corp., San Jose, CA) that has the capability of controlling the acceleration very precisely. The sensor was exposed to sinusoidal forces at different frequencies. At each frequency, the maximum acceleration was adjusted such that the maximum capacitance C_{max} measured was approximately 1 μF. The acceleration, as calculated from Eq. (1.38), was then plotted versus frequency along with the actual measured value of the acceleration (to confirm the measurements, another commercially available sensor, Model 7600B1 from Dytran Instruments, Inc., Chatsworth, CA, was mounted on the electrodynamic shaker together with the present sensor). The result is shown in Figure 1.32.

Concerning the value of the damping coefficient β, it was calculated from Eq. (1.39) and by observing the maximum acceleration a at the resonant frequency (or natural frequency) ω_0. Based on this measurement, a simple calculation showed that $\beta \approx 0.8$ for the present sensor. Figures 1.33 and 1.34 show the measured acceleration versus frequency for a maximum measured capacitance of 1 nF and 1pF, respectively. As expected, an acceleration as little as 10^{-13}g can be easily detected with the sensor.

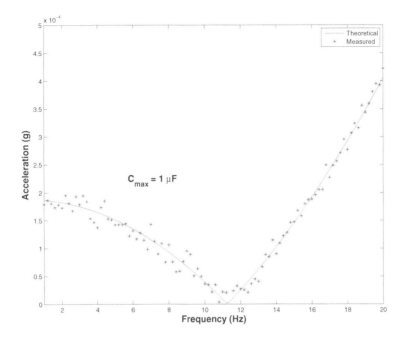

Figure 1.32: Measured acceleration versus frequency for a peak capacitance of 1 μF.

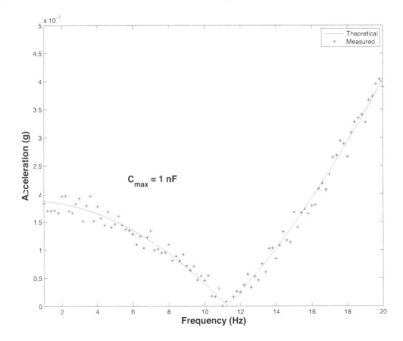

Figure 1.33: Measured acceleration versus frequency for a peak capacitance of 1 nF.

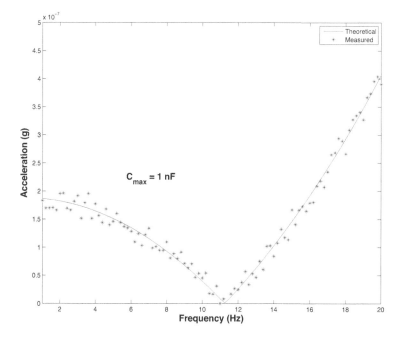

Figure 1.34: Measured acceleration versus frequency for a peak capacitance of 1 pF.

1.3.4 Conclusion

The new ultrahigh sensitivity pressure and vibration sensor described in this section is sensitive to pressures as little as 14 nanopascal and to accelerations as little as 10^{-13}g. It is therefore extremely useful in demanding new applications such as earthquake prediction and the measurement of pressure in biological processes. The prototype device that was fabricated and tested exhibited a change in capacitance of approximately 54 μF over a pressure range of 0.77 Pa. The equations that relate the applied pressure or acceleration to the measured capacitance of the device are fairly simple, and testing has shown a very good agreement between theory and experiment. It should be pointed out that the sensor must be mounted essentially in the vertical position, as shown in the figures; however, tilt angles of up to 45° will not affect the capacitance measured. Larger tilt angles, however, will result in the wetting of a portion of the surface of the upper electrode and can result in a degradation of the accuracy of the sensor.

1.4 Quiz

1) True or false: novel new microscale capacitive pressure sensors are based on the displacement of a plate.

Ans: False. These new sensors are based on the deformation of a droplet of mercury.

2) True or false: both capacitive and inductive pressure sensors are linear.
Ans: False. Only the inductive pressure sensors are linear. Capacitive pressure sensors are usually nonlinear.

3) Which scientific principle is behind ultraminiature, ultrasensitive pressure sensors?
Ans: The use of a variable ultracapacitor.

References

[1] C.S. Sander, J.W. Knutti, and J.D. Meindl, A monolithic capacitive pressure sensor with pulse-period output, *IEEE Trans. on Electron Devices*, vol. 27, 5, pp. 927–930, 1980.

[2] H. Chau and K.D. Wise, An ultraminiature solid state pressure sensor for a cardiovascular catheter, *IEEE Trans. on Electron Devices*, vol. 35, 12, pp. 2355–2362, 1988.

[3] A. Yasukawa, M. Shimazoe, and Y. Matsuoka, Simulation of circular silicon pressure sensors with a center boss for very low pressure measurement, *IEEE Trans. on Electron Devices*, vol. 36, 7, pp. 1295–1302, 1989.

[4] J.T. Kung and H.S. Lee, An integrated air-gap capacitor pressure sensor and digital readout with sub-100 attofarad resolution, *IEEE J. of Microelectromechanical Systems*, vol. 1, 3, pp. 121–129, 1992.

[5] C.H. Mastrangelo, X. Zhang, and W.C. Tang, Surface micromachined capacitive differential pressure sensor with lithographically defined silicon diaphragm, *IEEE J. of Microelectromechanical Systems*, vol. 5, 2, pp. 98–105, 1996.

[6] A.V. Chavan and K.D. Wise, Batch-processed vacuum-sealed capacitive pressure sensors, *IEEE J. of Microelectromechanical Systems*, vol. 10, 4, pp. 580–588, 2001.

[7] M.X. Zhou, Q.A. Huang, M. Quin, and W. Zhou, A novel capacitive pressure sensor based on sandwich structures, *IEEE J. of Microelectromechanical Systems*, vol. 14, 6, pp. 1272–1282, 2005.

[8] J.N. Palasagaram and R. Ramadoss, MEMS capacitive pressure sensor fabricated using printed circuit processing techniques, *IEEE Sensors Journal*, vol. 6, 6, pp. 1374–1375, 2006.

[9] E.G. Bakhoum and M.H.M. Cheng, Capacitive pressure sensor with very large dynamic range, *IEEE Trans. on Components and Packaging Technology*, vol. 33, 1, pp. 79–83, 2010.

[10] S.A. Dyer, *Survey of Instrumentation and Measurement*, Wiley, New York, 2001.

[11] W.C. Dunn, *Fundamentals of Industrial Instrumentation and Process Control*, Artech House Publishers, Boston, 2005.

[12] Kalvico Corp., *Pressure Sensors*, http://www.kavlico.com/products/transportation/pressure.php, 2009.

[13] Servoflo Corp., *SM5470 Pressure Transducers*, http://www.servoflo.com/pressure-sensor-overview/servoflo-pressure-sensors/smi-pressure-sensors/sm5470.html, 2009.

[14] Setra Corp., *Capacitive Pressure Sensors*, http://www.setra.com/tra/pro/pdf/264.pdf, 2009.

[15] O.O. Van der Biest and L.J. Vandeperre, Electrophoretic deposition of materials, *Annu. Rev. Mater. Sci.*, 29, pp. 327–352, 1999.

[16] T. Remmel et al., Barium strontium titanate thin film analysis, *Advances in X-ray Analysis*, vol. 43, pp. 510–518, 2000.

[17] S. Oh, J.H. Park, and J. Akedo, Dielectric characteristics of barium strontium titanate films prepared by aerosol deposition on a Cu substrate, *IEEE Trans. on Ultrasonics, Ferroelectrics, and Frequency Control*, vol. 56, 3, pp. 421–424, 2009.

[18] E.A. Parr, *IC 555 Projects*, Babani Publishing, Ltd., London, UK, p. 13, 1981.

[19] E. Matijevic and M. Borkovec, *Surface and Colloid Science*, Wiley, New York, 1969.

[20] E. Pisler and T. Adhikari, Numerical calculation of mutual capacitance between two equal metal spheres, *Physica Scripta*, vol. 2, 3, pp. 81–84, 1970.

[21] P.M. Fishbane, S.G. Gasiorowicz, and S.T. Thornton, *Physics for Scientists and Engineers*, 3rd Ed., Prentice-Hall, Englewood Cliffs, NJ, 2005.

[22] C.M. Harris and A.G. Peirsol, *Shock and Vibration Handbook*, McGraw Hill, New York, 2001.

[23] G.J. Radosavljevic, L.D. Zivanov, W. Smetana, A.M. Maric, M. Unger, and L.F. Nad, A wireless embedded resonant pressure sensor fabricated in the standard LTCC technology, *IEEE Sensors Journal*, vol. 9, 12, pp. 1956-1962, 2009.

[24] J. Wang, Z. Tang, J. Li, and F. Zhang, A micropirani pressure sensor based on the tungsten microhotplate in a standard CMOS process, *IEEE Transactions on Industrial Electronics*, vol. 56, 4, pp. 1086–1091, 2009.

[25] L.T. Ee, B.D. Pereles, and K.G. Ong, A wireless embedded sensor based on magnetic higher order harmonic fields: Application to liquid pressure monitoring, *IEEE Sensors Journal*, vol. 10, 6, pp. 1085–1090, 2010.

[26] J. Meyer, B. Arnrich, J. Schumm, and G. Troster, Design and modeling of a textile pressure sensor for sitting posture classification, *IEEE Sensors Journal*, vol. 10, 8, pp. 1391–1398, 2010.

[27] LVDT Pressure Sensors, Omega Corporation, Stamford, Connecticut, http://www.omega.com.

[28] Pressure Transducers, RDP Group, UK, http://www.rdpe.com

[29] LVDT—Linear Variable Differential Transducers, Honeywell Corp., Morristown, NJ, http://www.honeywell.com/sensing.

[30] B.S. Smith and A. Coull, *Tall Building Structures*, Wiley, New York, 1991.

[31] L. Knopoff et al., *Earthquake Prediction: The Scientific Challenge*, Proceedings of the National Academy of Sciences, vol. 93, 9, pp. 3719–3720, 1996.

[32] T. Takanami and G. Kitagawa, *Methods and Applications of Signal Processing in Seismic Network Operations*, Springer-Verlag, Berlin, 2003.

[33] P. Gasparini, G. Manfredi, and J. Zschau, *Earthquake Early Warning Systems*, Springer-Verlag, Berlin, 2007.

[34] A. Laudati, F. Mennella, M. Giordano, G. D'Altrui, C.C. Tassini, and A. Cusano, A fiber-optic bragg grating seismic sensor, *IEEE Photonics Technology Letters*, vol. 19, 24, pp. 1991–1993, 2007.

[35] S.G. Gevorgyan, V.S. Gevorgyan, H.G. Shirinyan, G.H. Karapetyan, and A.G. Sarkisyan, A radically new principle of operation seismic detector of nanoscale vibrations, *IEEE Trans. on Applied Superconductivity*, vol. 17, 2, pp. 1051–8223, 2007.

[36] E.M. Puchner, A. Alexandrovich, A.L. Kho, U. Hensen, L.V. Schfer, B. Brandmeier, F. Grater, H. Grubmuller, H.E. Gaub, and M. Gautel, Mechanoenzymatics of titin kinase, *Proceedings of the National Academy of Sciences*, vol. 105, pp. 13385–13390, 2008.

[37] G. Villanueva, J. Montserrat, F. Perez-Murano, G. Rius, and J. Bausells, Submicron piezoresistive cantilevers in a CMOS-compatible technology for intermolecular force detection, *Microelectronic Engineering*, vol. 73-74, pp. 480–486, 2004.

[38] E.G. Bakhoum and M.H.M. Cheng, Capacitive pressure sensor with very large dynamic range, *IEEE Trans. on Components and Packaging Technology*, vol. 33, 1, pp. 79–83, 2010.

[39] J. Schindall, The charge of the ultracapacitors, *IEEE Spectrum*, pp. 42–46, 2007.

[40] W. Hu, L. Yuan, Z. Chen, D. Gong, and K. Saitob, Fabrication and characterization of vertically aligned carbon nanotubes on silicon substrates using porous alumina nanotemplates, *J. of Nanoscience and Nanotechnology*, vol. 2, 2, pp. 203–207, 2002.

[41] B.E. Conway, *Electrochemical Supercapacitors*, Kluwer Academic Publishers, New York, 1999.

[42] A. Burke, Ultracapacitors: why, how, and where is the technology, *J. of Power Sources*, vol. 91, pp. 37–50, 2000.

[43] E.G. Bakhoum, New Mega-Farad Ultracapacitors, *IEEE Trans. on Ultrasonics, Ferroelectrics, and Frequency Control*, vol. 56, 1, pp. 14–21, 2009.

[44] R. Feynman et al., *The Feynman Lectures on Physics*, vol. 1, Addison Wesley, Reading, MA, 1964.

[45] P.A. Tipler, *Physics*, Worth Publishers, New York, 1986.

[46] P.H. Sydenham, Acceleration Measurement, in *Handbook of Measuring System Design*, Wiley, New York, 2005.

[47] P. Gardonio, M. Gavagni, and A. Bagolini, Seismic velocity sensor with an internal sky-hook damping feedback loop, *IEEE Sensors Journal*, vol. 8, 11, pp. 1776–1784, 2008.

[48] W. Boyes, *Instrumentation Reference Book*, Butterworh-Heinemann/Elsevier, Burlington, MA, 2010.

[49] J.E. Holmes, D. Pearce and T.W. Button, Novel piezoelectric structures for sensor applications, *Journal of the European Ceramic Society*, vol. 20, 16, pp. 2701–2704, 2000.

[50] E.A. Parr, *IC 555 Projects*, Babani Publishing, Ltd., London, UK, p. 13, 1981.

Chapter 2

Motion and Acceleration Sensors

2.1 Ultrahigh Sensitivity, Wide Dynamic Range Sensors

A number of new applications, such as self-guided small projectiles and autonomous surveillance airplanes, have created a requirement for a highly sensitive yet very small acceleration sensor. The fundamental problem in acceleration sensors is the size–sensitivity tradeoff [1–4]. Specifically, a highly sensitive acceleration sensor must necessarily be a large sensor [1–4]. A new miniature acceleration sensor with a radical new design was recently introduced. The sensor offers a very large dynamic range and high sensitivity. This new type of acceleration sensor will be available commercially within a time span of 1–2 years.

2.1.1 Structure

The new acceleration sensor is based on the concept of creating a variable ultracapacitor structure that consists of one small droplet of electrolyte that is positioned between two carbon nanotube (CNT) electrodes. At rest, the CNT electrodes remain outside of the electrolyte due to their hydrophobic nature. Under acceleration, however, the inertial forces push the CNT electrodes into the electrolyte and the typical capacitance of an ultracapacitor is obtained. The prototype described here has shown a capacitance variation of approximately 5 μF under an acceleration increase from zero to 2200 g. The sensitivity is therefore 2.27 nF/g. The concept behind the new sensor is shown in Figure 2.1.

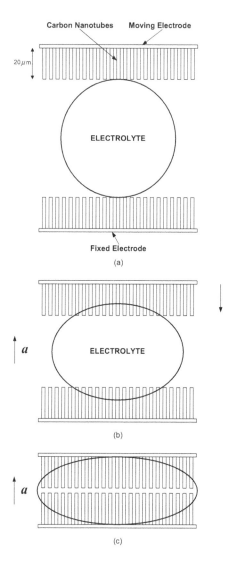

Figure 2.1: Principle of operation of the new acceleration sensor (dimensions greatly exaggerated).

As shown in Figure 2.1(a), a small droplet of electrolyte (fluid with high ionic conductivity) is placed between two electrodes on which CNTs of a length of 20 μm are grown. One electrode is fixed while the other is movable, as shown. Since the weights of the droplet and the electrodes are very small, and since CNTs are superhydrophobic, the CNTs do not penetrate the electrolyte when the mechanism is at rest. The capacitance between the two electrodes is hence approximately equal to

Figure 2.2: Photograph of the sensor. The SOIC package measures 6 mm × 8 mm. The diameter of the electrolyte droplet is 2 mm, and the diameter of the moving electrode is 3 mm.

zero in that configuration. In Figure 2.1(b), as acceleration occurs along the axis of the device (here, vertically, upward), the inertial forces created by the droplet and the moving electrode cause the CNTs to penetrate the electrolyte, as shown. Accordingly, the capacitance between the two electrodes (which is proportional to the immersed area of the CNTs) increases. In Figure 2.1(c), the forces created due to inertia further increase at higher accelerations, and the penetration of the CNTs into the electrolyte increase to a maximum. The capacitance of the thus-formed ultracapacitor increases to a substantially high value (approximately 5 μF in the present prototype).

It is to be pointed out that the mechanism shown in Figure 2.1 is essentially two ultracapacitors in series, as capacitance forms at the interface between the electrolyte and the surface of the CNTs (see [5,6] for an introduction to ultracapacitor technology). Figure 2.2 shows a photograph of the actual prototype that was assembled by the author. As the figure shows, an SOIC chip that measures 6 mm × 8 mm has an open cavity in which the mechanism is embedded. The figure shows the electrolyte droplet (with a diameter of 2 mm) and the upper electrode (removed and placed next to the chip). The lower electrode is underneath the droplet and is not shown in the figure.

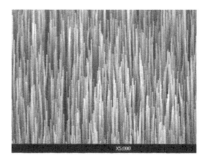

Figure 2.3: Multiwalled carbon nanotubes of an average diameter of 250 nm and a length of about 20 μm, grown on a stainless steel electrode.

The sensor, as described, is a one-axis sensor, and can be used to measure acceleration along the X, Y, or Z axes. Of course, if a 3-axis acceleration sensor is desired, then 3 units of the sensor can be packaged and used together. The electrodes used in the sensor are stainless steel electrodes, where, as indicated, a layer of CNTs is grown on one side of each electrode. These CNTs were deposited by a specialized commercial supplier that utilizes the process of chemical vapor deposition [10, 11]. Figure 2.3 is a scanning electron microscope (SEM) micrograph that shows the CNT layer.

Figure 2.4 shows a mechanical diagram of the main components of the sensor. As the figure shows, the electrolyte droplet, the fixed electrode, and the moving electrode are placed inside the cavity in the SOIC chip. A very thin aluminum cylinder with a height of approximately one-half that of the cavity is also inserted in the

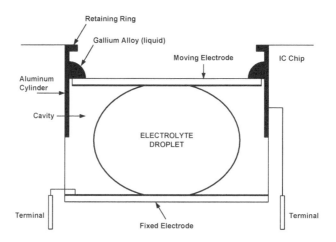

Figure 2.4: Mechanical structure of the sensor (scale = 16:1).

Table 2.1 **Comparison of the new sensor to other known types of capacitive acceleration sensors.**

	Sensitivity	Dynamic Range	Size
New sensor	2.27 nF/g	2200 g	3 mm (dia) x 2 mm (h)
Other capacitive acceleration sensors	typically a few pF/g	typically a function of the size	few cm up to a few hundred cm (for each dimension)

cavity. The purpose of the cylinder is to act as a contact terminal for the moving electrode. As shown in the figure, a small trace of a conductive liquid that is known commercially as "Liquid Metal" is placed on top of the moving electrode and serves to establish electrical contact between the electrode and the aluminum cylinder. The aluminum cylinder is, in turn, connected to an external terminal. The "Liquid Metal" is a gallium alloy that is normally liquid at temperatures above $-20°C$ [12]. As also shown in Figure 2.4, a steel ring is placed on top of the cavity and serves to retain the moving electrode, and hence the electrolyte droplet, inside the cavity (the retaining ring appears in Figure 2.2). The electrolyte used in the present application is propylene carbonate (solvent) in which an ionic salt is dissolved (a typical ultracapacitor electrolyte [5, 6]).

Table 2.1 shows a comparison of this sensor to other capacitive acceleration sensors that exist in the literature (see [7, 8]). In addition, the author would like to bring to the attention of the reader an earlier publication in which the concept of using a variable ultracapacitor structure to sense pressure was shown [9].

Two important observations must now be made. First, it is to be pointed out that propylene carbonate has a boiling point of 240°C. Accordingly, any possible evaporation of the electrolyte will occur only at elevated temperatures. The sensor was indeed tested at temperatures of up to 80°C, and no evaporation of any kind was observed. The second issue concerns the possible effect of the inclination angle on the measurement provided by the sensor. It is to be pointed out that the weight of the electrolyte droplet is only 5 mg, and hence the surface tension forces are far larger than the deformation forces that exist due to the weight. Furthermore, the droplet is confined within a restricted space, as shown in Figure 2.4, and therefore it is always in contact with the CNT electrodes (even if it shifts laterally). These facts indicate that the inclination angle has absolutely no effect on the measurement provided by the sensor. This was indeed confirmed through the experiments.

2.1.2 Theory

Figure 2.5 depicts an array of CNTs that is penetrating a liquid electrolyte. For each CNT, the hydrophobic force (or force that repels the fluid away from the CNT) will be given by

$$F = 2\pi r\gamma\cos(180° - \theta) \tag{2.1}$$

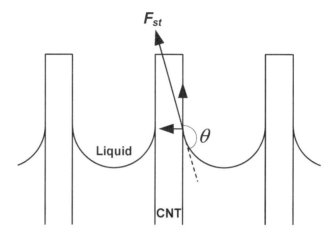

Figure 2.5: Array of CNTs penetrating a liquid electrolyte.

where r is the radius of one CNT, γ is the surface tension of the fluid, θ is the liquid–solid contact angle, and the product $2\pi r\gamma$ represents the surface tension force [13,14].

The total upward force that is acting on the fluid will be given by NF, where N is the total number of CNTs penetrating the fluid. We shall first develop a formulation for the capacitance at the interface between the moving electrode and the electrolyte (see Figure 2.1). Given an electrode with a mass m that is moving with a uniform acceleration a, the Newtonian force $F = ma$ must be balanced by the total hydrophobic force, that is,

$$F_{total} = ma = 2\pi r N\gamma\cos(180° - \theta) \qquad (2.2)$$

The surface area A of the CNTs that is immersed in the electrolyte will be given by

$$A = 2\pi r \sum_{i=1}^{N} x_i \qquad (2.3)$$

where x_i is the immersion depth of any given CNT. The capacitance of the electrode–electrolyte interface will be now given by [15]

$$C = \frac{\varepsilon_0\varepsilon_r A}{d} \qquad (2.4)$$

where ε_0 is the permittivity of free space, ε_r is the relative permittivity of the electrolyte,[1] and d is the separation between the positive and negative charges at the interface (this separation is about 1 nm in ultracapacitors [5,6]). The above expression is the well-known equation of the capacitance of a parallel-plate capacitor. It is

---

[1]For ultracapacitors, the relative permittivity of the 1 nm thick electrolyte layer at the interface is approximately equal to 1.

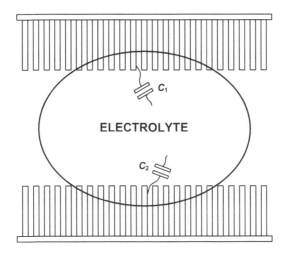

Figure 2.6: The total capacitance of the ultracapacitor is the series combination of the capacitances at the two interfaces C_1 and C_2.

important to note here that the capacitance of a cylindrical capacitor is very nearly the same as the capacitance of a parallel-plate capacitor when the interelectrode separation is much smaller than the radius of the cylinder (or the CNT in this application). From the previous two equations, and by taking $\varepsilon_r \approx 1$, we have

$$C = \frac{\varepsilon_0}{d}\left(2\pi r \sum_{i=1}^{N} x_i\right) \tag{2.5}$$

In the present application, the capacitance between the moving electrode and the electrolyte is not directly measurable, however. What is measurable, in fact, is the series combination of the capacitances C_1 and C_2 at the two interfaces [5,6]. This is the observed capacitance of the ultracapacitor (Figure 2.6).

In general, C_2 will be larger than C_1, since the lower interface is acted upon by both the upper electrode and the electrolyte droplet. Now, given the general form of Eq. (2.5), the series combination of C_1 and C_2 can be expressed as

$$C_{total} = K\frac{\varepsilon_0}{d}\left(2\pi r \sum_{i=1}^{N} x_i\right) \tag{2.6}$$

where K is a constant that ranges from 1 at very small accelerations (where $C_2 \gg C_1$) to 0.5 at very high accelerations (where $C_2 \approx C_1$). From Eqs. (2.2) and (2.6) we now have

$$ma = N\gamma\cos(180° - \theta)\left(\frac{dC_{total}}{K\varepsilon_0 \sum_{i=1}^{N} x_i}\right) \tag{2.7}$$

Hence,

$$a = -\frac{dN\gamma\cos\theta}{Km\varepsilon_0 \sum_{i=1}^{N} x_i}C_{total}, \quad \theta > 90° \tag{2.8}$$

Equation (2.8) can be greatly simplified by observing the following fact: the length of the CNTs is very short (20 μm, or less than the thickness of a human hair). Then, to a very good approximation, $\sum_{i=1}^{N} x_i \approx Nx$, where $x = 20 \mu$m is the full length of the CNTs (in other words, the variable is the number N of CNTs, while the immersed depth is assumed to be constant). Equation (2.8) now simplifies as

$$a = -\frac{d\gamma\cos\theta}{Km\varepsilon_0 x} C_{total}, \quad \theta > 90° \tag{2.9}$$

It is to be pointed out that the length of the CNTs is extremely small (20 μm), and hence no "threshold" for the penetration was observed in the operation of the sensor. As pointed out, the operation depends mainly on the droplet's spreading rather than on the droplet's penetration into the CNTs. The only purpose of using CNTs in this application is to provide an extremely large surface area and hence a very large capacitance.

2.1.3   Experimental Results

Physical parameters of the sensor and the dynamic range:
The following are the physical parameters that characterize the present sensor:

■ $d \approx 1$ nm (see [5, 6])

■ $\gamma = 41 \times 10^{-3}$ N/m [16]

■ $\theta \approx 160°$ (according to a number of published physical studies on the wetting of CNTs by organic liquids similar to propylene carbonate [17, 18]. It is important here to point out that the force equation can be written in terms of $\cos\theta$ or $\cos(180° - \theta)$. The contact angle θ will be substantially different in the two cases)

■ $m = 0.1$ g

■ $x = 20 \mu$m

■ Area of the moving electrode $= 0.071$ cm^2

■ $N = 10^9$ CNTs per cm^2; hence N on the electrode $= 71 \times 10^6$ CNTs

■ r (radius of one CNT) ≈ 125 nm

By substituting with the above parameters in Eq. (2.9), we obtain

$$a = 2.18 \times 10^9 \frac{C_{total}}{K} \quad \text{m/s}^2 \tag{2.10}$$

Substitution in Eq. (2.5) and assuming full use of the CNTs (i.e., $N = 71 \times 10^6$) gives the maximum capacitance per interface as 9.87 μF. Hence, the maximum capacitance of the device (C_{total}, max) is approximately 4.94 μF. With the constant

K ranging from 0.5 at high accelerations to approximately 1 at low accelerations, we can conclude from Eq. (2.10) that the maximum measurable acceleration will be approximately 21517 m/s², or 2193 g (a substantially high acceleration). Theoretically, the minimum acceleration that can be measured is 0; however, measurement of a capacitance that is less than 1 pF is not usually reliable. If we now assume that C_{total} (min) $\approx$ 1 pF and that $K \approx 1$, Eq. (2.10) shows that the minimum measurable acceleration is approximately 2×10^{-3} m/s² or 2×10^{-4} g. It is therefore clear that the dynamic range of this sensor is substantially large.

Measurement of the capacitance constant K:

The capacitance constant K was determined by mounting the sensor on an electrodynamic shaker [7, 8, 19, 20] and measuring the capacitance C_{total} as a function of acceleration. Three different types of measurements were performed: for very low accelerations (less than 1 g), an electrodynamic shaker from Data Physics Corp. (San Jose, CA) was used in the testing. For small to medium accelerations (1 g to 100 g), an electrodynamic shaker from Dynamic Solutions, Inc. (Northridge, CA) was used. For high accelerations (greater than 100 g) the sensor was actually tested by mounting the sensor and its associated circuitry inside a low-speed 0.45-caliber bullet (see further details below). When tested on the electrodynamic shaker, the sensor was subjected to shock pulses of a duration of about 0.2 s and different known accelerations. The acceleration in each case was also measured simultaneously with an off-the-shelf sensor from Freescale Semiconductors, Inc. (Austin, TX). To record the acceleration and the associated capacitance in each test, both physical variables were converted to signals and observed on digital storage oscilloscopes (see further details below). The value of the unknown constant K was then calculated from Eq. (2.10) for each measurement performed. The results are shown in Figure 2.7.

As Figure 2.7 shows, K does indeed vary linearly from approximately 1 at very low accelerations to approximately 0.5 for accelerations above 2000 g. As the figure also shows, for accelerations that are less than about 100 g, K is very nearly equal to

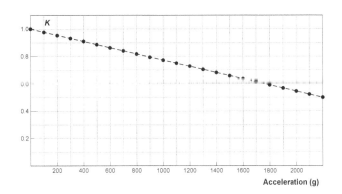

Figure 2.7: The capacitance constant K as a function of acceleration (0.0002 g to 2200 g).

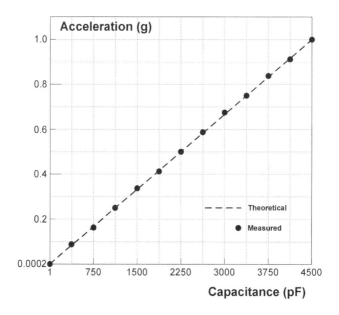

Figure 2.8: Measured and theoretical acceleration in the range of 0.0002 g to 1 g as a function of the measured capacitance.

1, and hence the acceleration a in Eq. (2.10) will be simply proportional to C_{total} for that wide range of accelerations.

Measurement of acceleration:

By using the same experimental setup, Eq. (2.10) was tested directly by measuring the capacitance of the sensor during the acceleration pulse and calculating the value of the acceleration predicted by the equation. The value of the acceleration was also measured simultaneously with the capacitance by using an off-the-shelf sensor. The range of the accelerations used in this test was the low range of 2×10^{-4} g to 1 g, and K was taken to be equal to 1. The results are shown in Figure 2.8. Clearly, the sensor is quite linear in that range.

The sensor was subsequently tested in the small to medium range of accelerations (1 g to 100 g). Figure 2.9 shows the actual acceleration that was measured versus the value predicted by Eq. (2.10) without correction for the value of K. The maximum error is approximately 3%, near the higher values of acceleration. Of course, if corrected for the variation in K, the acceleration predicted by Eq. (2.10) would precisely match the actual value.

The sensor was finally tested in the acceleration range of 100 g to 2200 g. No mechanical shaker is capable of achieving such high values of acceleration, and for that reason the sensor was tested by actually mounting it inside a 0.45-caliber bullet. Figure 2.10 shows the type of bullet that was used, the circuit board on which the sensor was mounted, the 1.5 V battery that was used for powering the board, and the

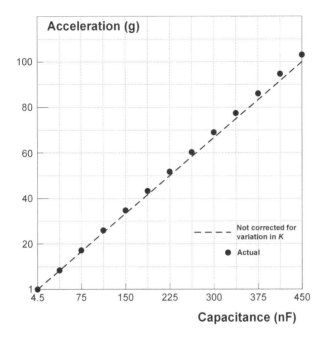

Figure 2.9: Acceleration in the range of 1 g to 100 g as a function of the measured capacitance. Without correction for the value of K, the maximum error is approximately 3%.

Figure 2.10: The circuit board and the 0.45-caliber bullet that were used for testing the sensor at accelerations up to 2200 g.

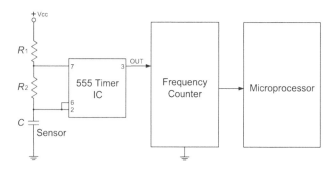

Figure 2.11: Block diagram of the interface circuit used for measuring the capacitance *C* of the sensor in real time.

handgun that was used for firing the bullets. It is to be pointed out that the acceleration of a normal bullet inside a gun's barrel is typically on the order of 10^6 g or higher. To get the lower range of 100 g to 2200 g that is applicable to the sensor, the chemical propellant inside the bullet's case was reduced considerably.

On the circuit board shown in Figure 2.10, a second, off-the-shelf acceleration sensor is also present (for confirmation of the acceleration calculation).[2] A block diagram of the circuit used is shown in Figure 2.11. As shown, the sensor (represented by the capacitance *C*), is connected to a well-known 555 oscillator circuit that generates a square wave with a frequency that is dependent on the capacitance (the circuit is described in references such as [21]). A frequency counter then converts the frequency to a digital readout. The digital readout is finally processed by an on-board 8 bit microprocessor. The microprocessor simply tracks the highest number and stores that number in a nonvolatile memory. In the circuit shown in Figure 2.10, the microprocessor actually processes two signals: one from the new sensor described here and the other from the off-the-shelf sensor that is also present on board.

The testing results for the acceleration range of 100 g to 2200 g are shown in Figure 2.12. As expected, the correct value of *K* must be used in Eq. (2.10), or otherwise serious errors will result in that high acceleration range.

2.1.4 Conclusion

The new MEMS acceleration sensor described in this chapter has an ultrawide dynamic rage and high sensitivity. Specifically, the sensitivity is 2.27 nF/g, and the dynamic range extends from 2×10^{-4} g to 2193 g. It will therefore be useful in a wide array of new applications, including self-guided projectiles and autonomous surveillance aircraft. The prototype device that was fabricated and tested has exhibited a change in capacitance of approximately 5 μF over the dynamic range of the

---

[2]It must be pointed out that no acceleration sensor that is available commercially matches the dynamic range and the sensitivity of the sensor described here.

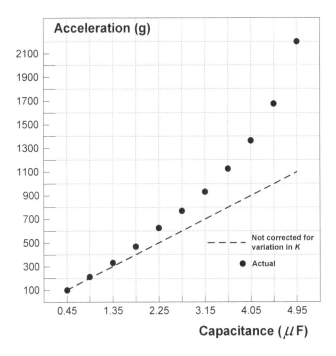

Figure 2.12: Acceleration in the range of 100 g to 2200 g as a function of the measured capacitance.

sensor. The sensor is linear for accelerations up to approximately 100 g and nonlinear for higher accelerations. In the nonlinear region, a capacitance constant K must be fully included in the acceleration–capacitance relationship to avoid large errors. The equation that relates the acceleration to the measured capacitance is fairly simple, and testing has shown a very good agreement between theory and experiment.

2.2 Other Motion and Acceleration Microsensors

Other approaches for motion and acceleration sensors have also been introduced recently. Figure 2.13 shows a concept for a monolithic integration of pressure plus acceleration composite Tire Pressure Monitoring Sensor (TPMS) with a single-sided micromachining technology. Figure 2.14 shows a photo of the actual sensor.

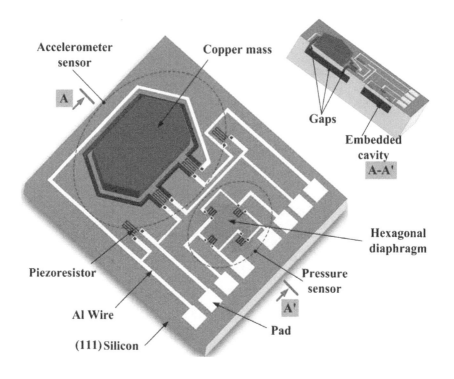

Figure 2.13: Concept for monolithic Tire Pressure Monitoring Sensor (TPMS), from [22].

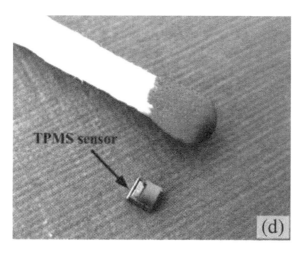

Figure 2.14: Photo of actual sensor, from [22].

2.3   Quiz

1) True or false: applications such as self-guided projectiles require an acceleration sensor with a large dynamic range and high sensitivity.
Ans: True.

2) Which scientific principle is behind the new ultraminiature, ultrasensitive acceleration sensor described in Chapter 2?
Ans: The use of a variable ultracapacitor.

References

[1] Quan Wang, Xinxin Li, Tie Li, Minhang Bao, and Wei Zhou, On-chip integration of acceleration, pressure, and temperature composite sensor with a single-sided micromachining technique", *IEEE J. of Microelectromechanical Systems*, vol. 20, 1, pp. 42–51, 2011.

[2] A. Vallan, M. Casalicchio, and G. Perrone, Displacement and acceleration measurements in vibration tests using a fiber optic sensor, *IEEE Trans. on Instrumentation and Measurement*, vol. 59, 5, pp. 1389–1398, 2010.

[3] Tran Due, Tan S. Roy, Nguyen Phu Thuy, and Huu Tue Huynh, Streamlining the Design of MEMS Devices: An Acceleration Sensor, *IEEE Circuits and Systems Magazine*, vol. 8, 1, pp. 18–21, 2008.

[4] L.P. Shen, K. Mohri, T. Uchiyama, and Y. Honkura, Sensitive acceleration sensor using amorphous wire SI element combined with CMOS IC multivibrator for environmental sensing, *IEEE Trans. on Magnetics*, vol. 36, 5, pp. 3667–3675, 2000.

[5] B.E. Conway, *Electrochemical Supercapacitors*, Kluwer Academic Publishers, New York, 1999.

[6] A. Burke, Ultracapacitors: why, how, and where is the technology, *J. of Power Sources*, vol. 91, pp. 37–50, 2000.

[7] S.A. Dyer, *Survey of Instrumentation and Measurement*, Wiley, New York, 2001.

[8] W.C. Dunn, *Fundamentals of Industrial Instrumentation and Process Control*, Artech House Publishers, Boston, 2005.

[9] E. Bakhoum and M.H. Cheng, Ultrahigh-sensitivity pressure and vibration sensor, *IEEE Sensors Journal*, vol. 11, 12, pp. 3288–3294, 2011.

[10] J. Schindall, The charge of the ultracapacitors, *IEEE Spectrum*, pp. 42–46, 2007.

[11] W. Hu, L. Yuan, Z. Chen, D. Gong, and K. Saitob, Fabrication and characterization of vertically aligned carbon nanotubes on silicon substrates using porous alumina nanotemplates, *J. of Nanoscience and Nanotechnology*, vol. 2, 2, pp. 203–207, 2002.

[12] Scitoys.com, *Liquid Metal*, http://scitoys.com/scitoys/scitoys/thermo/liquid-metal/liquid-metal.html, 2011.

[13] R. Feynman, R. Leighton, and M. Sands, *The Feynman Lectures on Physics*, vol.1, Addison Wesley, Reading, MA, 1964.

[14] P.A. Tipler, *Physics, Worth Publishers*, New York, 1986.

[15] W.H. Hayt and J.A. Buck, *Engineering Electromagnetics*, McGraw Hill, New York, 2006.

[16] A.W. Adamson and A. P. Gast, *Physical Chemistry of Surfaces*, Wiley, New York, 1997.

[17] A. Barber, S. Cohen, and H.D. Wagner, Static and dynamicwetting measurements of single carbon nanotubes", *Phys. Rev. Letters*, vol. 92, 18, Art. No. 186103, pp. 1–4, 2004.

[18] C. Journet, S. Moulinet, C. Ybert, S. T. Purcell, and L. Bocquet, Contact angle measurements on superhydrophobic carbon nanotube forests: Effect of fluid pressure, *EuroPhys. Letters*, vol. 71, 1, pp. 104–109, 2005.

[19] P.H. Sydenham, Acceleration Measurement, in *Handbook of Measuring System Design*, Wiley, New York, 2005.

[20] W. Boyes, *Instrumentation Reference Book*, Butterworh-Heinemann/Elsevier, Burlington, MA, 2010.

[21] E.A. Parr, *IC 555 Projects*, Babani Publishing Ltd., London, UK, p. 13, 1981.

[22] Jiachou Wang, Xiaoyuan Xia, and Xinxin Li, Monolithic integration of pressure plus acceleration composite TPMS sensors with a single-sided micromachining technology, *IEEE J. of Microelectromechanical Systems*, vol. 21, 2, pp. 284–293, 2012.

Chapter 3

Gas and Smoke Sensors

Gas and smoke sensors have also benefited greatly from the recent advances in nanotechnology. This chapter introduces a carbon monoxide (CO) gas detector that uses gold nanoparticles deposited on an array of carbon nanotubes as a sensing element, in addition to a smoke detector that achieves miniaturization by replacing a bulky alpha particle detection chamber with a silicon nanostructure. These types of detectors will be available commercially within a time span of 1 to 2 years.

3.1   A CO Gas Sensor Based on Nanotechnology

Carbon monoxide (CO) detectors are used extensively in the household and the industrial sector. Usually, a CO concentration in excess of 100 parts-per-million (ppm) is considered to be harmful for humans [1]. To detect such a small concentration, detectors have traditionally been large devices. The necessary sensitivity is achieved by using a large electrochemical cell or a group of heated semiconductor wires [2], where in either case the oxidation reaction of CO (to form CO_2) results in a small electric potential that can be detected. Very recently, a miniature, highly sensitive CO gas sensor that makes use of the very high catalytic activity of gold nanoparticles was introduced. Gold nanoparticles are deposited on the surfaces of an array of carbon nanotubes (CNTs). The CO gas is transformed into CO_2 as it comes in contact with the nanoparticles, and free electrons are released. The extra free electrons result in a measurable reduction in the resistivity of the CNT array. The detector has proved to be very sensitive at room temperature to CO gas concentrations as low as 100 ppm. It is characterized by a very small size and can be easily powered with a battery.

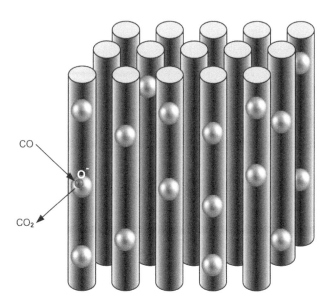

Figure 3.1: Fundamental principle of operation of the new CO detector.

3.1.1 Structure

The fundamental principle behind the new CO gas sensor is shown in Figure 3.1. As the figure shows, gold nanoparticles are attached to an array of multiwalled carbon nanotubes. Gold nanoparticles have a large affinity for oxygen, and oxygen molecules (from the ambient atmosphere) become adsorbed on the large surface presented by the nanoparticles, where they easily disassociate. CO, a reducing gas, easily combines with atomic (or molecular) oxygen and releases free electrons according to a well-known reaction. The presence of extra free electrons reduces the overall resistivity of the carbon nanotube array, which can be detected with an external circuit.

As shown in [3], [4], and other references, the two reactions that can be synthesized with catalysts are

$$
\begin{aligned}
2CO + O_2^- &\rightarrow 2CO_2 + e^- \\
CO + O^- &\rightarrow CO_2 + e^-
\end{aligned}
\tag{3.1}
$$

According to the principle shown in Figure 3.1, these well-known chemical reactions are synthesized in the new solid-state detector by using the excellent catalytic activity of gold nanoparticles [4] (while the high catalytic activity of gold nanoparticles has been known to chemists for several years, the application and the device described in Figure 3.1 are not shown in the engineering literature). Recent density functional theory studies have shown that such reactions have a barrier of less than 0.4 eV when a gold catalyst is used and are therefore synthesizable at temperatures below room temperature [4, 5].

The miniature solid-state CO detector described in this section has the following characteristic advantages:

■ Sensitivity to CO concentrations of less than 100 ppm at room temperature

■ Can be battery operated, with a very long battery life (the power required is quite negligible when the audio alarm is not activated)

■ Being a miniature detector, it can be placed virtually anywhere

The new technology reported here can be understood in proper context when compared with other CO sensing technologies that have appeared in the literature during the past decade. There are essentially two groups of new technologies. The first group, which requires temperatures significantly higher than room temperature, includes technologies such as field effect transistors and Schottky diodes that are coated with various metal oxides [7–12], sensors based on cobalt and rubidium catalysts [13, 14], sensors based on combinations of catalysts and polymers [15], and infrared detection techniques [16, 17]. The second group of technologies offers operation at room temperature in addition to high sensitivity. It includes optical techniques (specifically, absorption spectroscopy of CO) [18–20] and microwave techniques (specifically, rotational molecular resonance of CO molecules) [21]. This latter group of technologies, however, has the disadvantage of very high cost and the requirement for bulky detection equipment.

As far as the author is aware, the new technology reported here is the first technology that offers the combined features of high sensitivity, relatively low cost, very small size, and operation at room temperature.

3.1.2 Theory

The rate of the catalytic reactions in (3.1) can be described by the well-known reaction rate equation [22, 23]

$$-\frac{dP}{dt} = k_0 \exp\left(-\frac{E_a}{RT}\right) P^m \tag{3.2}$$

where P is the partial gas pressure, k_0 is the pre-exponential factor, E_a is the activation energy, and m is the pseudo-kinetic order of the reaction (the catalytic oxidation of CO by various catalysts shows a pseudo-kinetic order m that is usually 0 or 1 [24, 25]). A useful measure of the performance of a gas sensor is the conversion percentage (which, in this application, is the percentage of CO molecules that are converted to CO_2). The conversion percentage is usually expressed as [2]

$$\alpha(t) = 1 - \frac{P(t)}{P_0} \tag{3.3}$$

where P_0 is the partial pressure of CO at time $t = 0$. If $m = 1$, the solution of Eq. (3.2) is

$$P = P_0 \exp\left(-k_0 \exp\left(-\frac{E_a}{RT}\right) t\right), \qquad m = 1 \tag{3.4}$$

and hence the conversion percentage α is found to be

$$\alpha = 1 - \exp\left(-k_0 \exp\left(-\frac{E_a}{RT}\right) t\right), \qquad m = 1 \tag{3.5}$$

If $m = 0$, however, the solution of Eq. (3.2) is

$$P = P_0 - k_0 \exp\left(-\frac{E_a}{RT}\right) t, \qquad m = 0 \tag{3.6}$$

and α is found to be

$$\alpha = \frac{k_0}{P_0} \exp\left(-\frac{E_a}{RT}\right) t, \qquad m = 0 \tag{3.7}$$

Hence, a measurement of the conversion percentage of the reaction as a function of time and a determination of whether the rate is exponential as in Eq. (3.5) or linear as in Eq. (3.7) can lead to the determination of the value of m conclusively (note that Eq. (3.7) is a linear equation in time). Furthermore, a numerical solution of either Eq. (3.5) or Eq. (3.7) based on the data provided by such a measurement can lead to the determination of the activation energy E_a of the reaction. Most important, a measurement of the conversion percentage of the reaction, together with resistivity measurements, can lead to the characterization of the performance of the sensor at room temperature and at other temperatures.

3.1.3 Assembly of the Sensor

Multiwalled carbon nanotubes (CNTs) were obtained from a commercial supplier. The CNTs used have a diameter of approximately 50 nm, a length of approximately 20 μm, and were grown on a sheet of stainless steel. Two different methods were followed to attach gold nanoparticles to the CNTs: (1) electrodeposition, and (2) electroless deposition by means of a spontaneous redox reaction of gold ions in a solution [6]. Figure 3.2 shows a scanning electron microscope (SEM) photograph of gold nanoparticles attached to the lateral surfaces of the CNTs after the electrodeposition process. The electrodeposition was carried out with the CNTs acting as the cathode and with an electrolyte consisting of an $HAuCl_4$ (Au^{3+}) solution. Electric field intensities ranging from 5 V/cm to 100 V/cm were used. A lower field intensity and a longer exposure time generally resulted in better coatings and more uniform particle size distributions (the size of the deposited gold nanoparticles in this process ranged from 10 nm to 50 nm).

Figure 3.3 shows an SEM photograph of gold nanoparticles deposited with the direct redox reaction process. By comparing Figures 3.2 and 3.3, it is clear that the second process results in a substantially better coating and substantially more uniform particle size distribution. As discussed in [6], the deposition process was performed by immersing the CNTs in a solution of $HAuCl_4$ dissolved in equal volumes of ethanol and distilled water, at a concentration of 5 mM. After the CNTs were

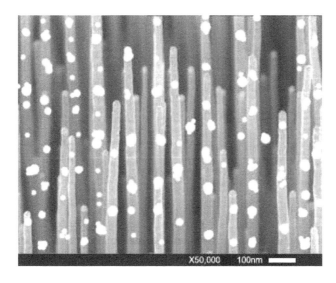

Figure 3.2: SEM micrograph of gold nanoparticles deposited on multiwalled carbon nanotubes by electrodeposition. The CNTs were exposed for 20 min under an electric field intensity of 5 V/cm.

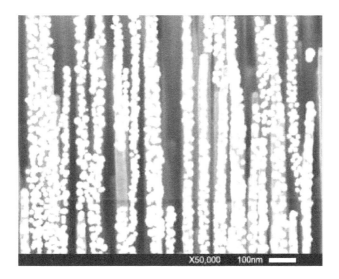

Figure 3.3: SEM micrograph of gold nanoparticles deposited on multiwalled carbon nanotubes by direct redox reaction.

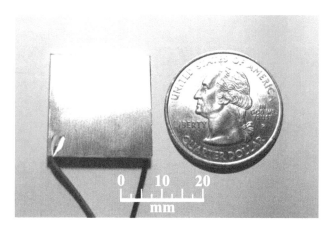

Figure 3.4: Photograph of the assembled detector (interface circuit not included).

immersed for a period of 3 min, they were rinsed with distilled water and dried in a nitrogen atmosphere.

The CNT array used in the present detector measures 20 mm × 20 mm. After the deposition of the gold nanoparticles, a second electrode was formed on top of the CNT array by sputtering. The second electrode is aluminum, and has sufficient porosity due to the deposition by sputtering. The pores in the aluminum electrode thus allow the CO gas to easily penetrate and reach the gold nanoparticles attached to the CNT array. The assembly of the steel electrode, the aluminum electrode, and the CNT array sandwiched between the two electrodes, constitutes the complete detector. The resistivity measurements described in Section 3.4 were taken between the two electrodes of the detector. Figure 3.4 shows a photograph of the finished detector.

3.1.4   Experimental Results

Measurement of the conversion percentage as a function of time and temperature:

Activity tests were performed to determine the conversion percentage of the reaction as a function of time, and for different values of temperature. The sensor was placed inside a small stainless steel reaction chamber with a volume of 3 cc, which is slightly larger than the volume of the sensor. The reaction chamber was in turn placed inside a programmable temperature container. CO gas mixed with air, at concentrations ranging from 100 ppm to 600 ppm, was then pumped into the reaction chamber. The rate of flow of the gas/air mixture was measured with a flowmeter and was adjusted to different values, the maximum rate being 3 cc/10 ms (hence, the measurements reported here have a resolution of 10 ms). The outlet of the chamber was connected to a hydrocarbon gas analyzer (model NGA 2000, from Emerson Process Management Corp.) for monitoring the residual concentration of CO after

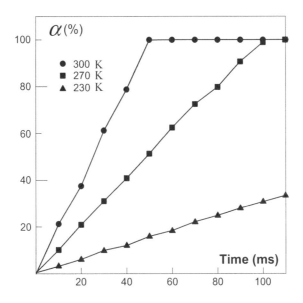

Figure 3.5: Conversion percentage as a function of time for temperatures ranging from 230 K to 300 K.

passage through the sensor. It should be pointed out that the air used in the tests was ordinary room air that was filtered with a charcoal filter in order to remove moisture and contaminants (more results related to the moisture content are given in the following section). Figure 3.5 shows three different plots of the conversion percentage α as a function of time. The tests were conducted at temperatures of 300 K, 270 K, and 230 K. The data in Figure 3.5 is associated with a CO concentration of 100 ppm.

It is clear that the plots in Figure 3.5 show a linear, rather than exponential, relationship. Accordingly, the value of the constant m is 0. It is also clear that total conversion (from CO to CO_2) occurs very rapidly at room temperature (within 50 ms). The value of the slope for any of the linear plots in Figure 3.5 is given by $k_0/P_0 \exp(-E_a/RT)$. By comparing the slopes of two different plots, the value of the ratio k_0/P_0 as well as of the activation energy E_a can be determined. From the data shown in the figure, E_a was estimated to be approximately 15 kJ.mol^{-1}. This value is in close agreement with earlier measurements that were reported in the literature [26, 27].

Figure 3.6 shows the effect of the CO concentration in the gas/air mixture. The measurements in Figure 3.6 were all taken at a fixed temperature of 300 K. Clearly, the effect of the concentration on the performance of the sensor is minimal. It can be noticed, however, that the reaction will require a slightly longer time for completion at higher concentrations, as it is obvious that more CO molecules will be competing for access to a fixed number of catalyst sites.

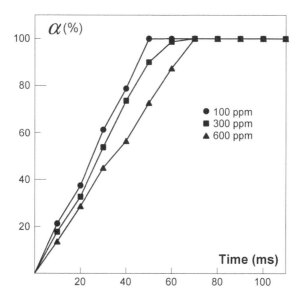

Figure 3.6: Conversion percentage as a function of time at a fixed temperature of 300 K and for 3 different CO gas concentrations.

Measurement of the resistance of the CNT array as a function of CO concentration:

The resistance of the CNT array was measured with a highly sensitive ohmmeter as the CO gas was repeatedly turned ON for a duration of 10 sec and OFF for a duration of 10 sec. Two multimeters were used to perform and verify the measurements: a specially modified Keithley model 610C multimeter, and an Agilent model 34420A multimeter. Both multimeters are capable of directly measuring a resistance of 100 nΩ or higher. For measurement of resistances that are less than 100 nΩ, however, the two mentioned multimeters were used in a 4-wire configuration, where one multimeter acts as a current source and the other as a voltmeter.[1] The first test was conducted with a CO gas concentration of 100 ppm. The result is shown in Figure 3.7. As shown, the resistance drops from a nominal value of 0.5 mΩ to approximately 1 $\mu\Omega$ when the gas is present. Hence, a good sensitivity to gas concentrations of 100 ppm or less is evident. The second test was conducted with a gas concentration of 600 ppm. The result is shown in Figure 3.8. It is evident that the response of the sensor is far greater, with the overall resistance dropping to an average value of 10 nΩ in that case.

It is to be pointed out that the area of the sensor (that is, of the CNT array) can be reduced to make the sensor smaller; however, the variation in the resistance of the

---

[1] Please note: even though the assembled sensor in Figure 3.4 is shown with only two wires, four wires were connected to the sensor in order to perform the measurements. The test itself is very well-known in the electronics literature and will not be described in detail here.

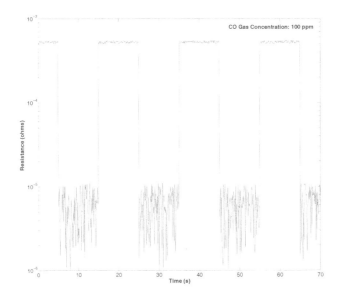

Figure 3.7: Resistance of the CNT array (log scale) as a function of time, as the CO gas (concentration: 100 ppm) is turned ON and OFF. The measurement was performed at room temperature. The nominal (upper) value of the resistance is 0.5 mΩ, and the lower value (measured when the gas was present) is approximately 1 $\mu\Omega$.

sensor assembly will be less in this case (as can be concluded from the well-known relationship between resistance and cross-sectional area).

3.1.5   *Auxiliary Experimental Results*

Effect of temperature on the performance of the sensor:

The effect of temperature on the reaction rate is shown in Figure 3.4. The reaction rate (and hence the sensor's response) is quite acceptable at room temperature, but can be much worse at substantially lower temperatures. The sensor was also tested at elevated temperatures (up to 80°C), and the reaction rate was found to be generally higher at such temperatures.

Effect of moisture on the performance of the sensor:

Moisture was added to the gas/air mixture (by circulating the mixture through a valve-controlled container filled with distilled water), and the relative humidity in the mixture was monitored with a moisture sensor. Figure 3.9 shows the result of a test that was conducted at room temperature and with a CO concentration of 100 ppm. As the figure shows, the effect of moisture was negligible for the relative humidity (RH) range of 5% to 40%. For the range of RH from 40% to 80%, however, the reaction rate slowed significantly. This can be understood on the basis of the fact that

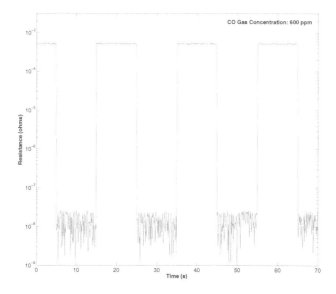

Figure 3.8: Resistance of the CNT array (log scale) as a function of time, as the CO gas (concentration: 600 ppm) is turned ON and OFF. The measurement was performed at room temperature. The nominal (upper) value of the resistance is 0.5 mΩ, and the lower value (measured when the gas was present) has an average value of 10 nΩ.

the hydrogen present in the moisture is itself a reducing gas and will compete with the CO molecules for access to the available catalyst sites.

Other performance parameters:

Tests to determine the repeatability of the measurements were conducted. For the same gas concentration, temperature, and relative humidity conditions, the measured resistance of the sensor was found to be generally repeatable within the bounds of the measurements shown in Figures 3.7 and 3.8.

Other tests were conducted to determine the sensitivity of the sensor to contaminants. Small amounts of contaminants such as hydrogen, ethanol, and benzene were introduced into the gas/air mixture flowing into the sensor. When the contaminant was a reducing gas, such as hydrogen, the effect on the performance of the sensor was generally proportional to the amount of contaminant present. Otherwise, the contaminant had no effect on the CO $\rightarrow$ CO_2 reaction rate.

3.1.6 *Conclusion*

The miniature, highly sensitive CO gas sensor presented in this section has the smallest size and the highest sensitivity at room temperature among all the known types of CO gas detectors. Furthermore, the detector requires only minimal power

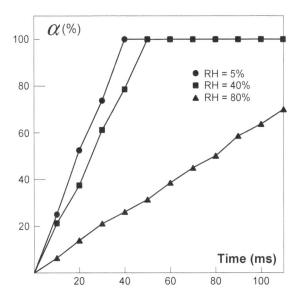

Figure 3.9: Effect of relative humidity on the reaction rate at room temperature (CO gas concentration = 100 ppm).

(in comparison with conventional CO sensors, which require heating of the sensing element) and can be easily powered with a battery. Possible circuits for interfacing this type of sensor to audio/visual alarming mechanisms can be found in references such as [20, 21]. The effective power consumption of these circuits is normally very low, and hence they are suitable for battery operation.

One disadvantage of this new sensor is the cost. CNT arrays are expensive to produce at the present time, and gold nanoparticles must be deposited on the CNT array (although the mass of gold used in the sensor is only about 1 μg). With mass production, however, the cost can be substantially reduced. Regardless, the sensor can be very useful in applications where a small size and high sensitivity are required.

3.2 Smoke Detectors

In this section we introduce a novel new smoke detector that is characterized by a very small size and very high sensitivity. The new detector is fundamentally based on an α-particle radiation source like the well-known α-particle smoke detector; however, instead of utilizing the principle of ionization of the air, the α particles are made to strike the gate of an n-channel MOSFET. This results in a net positive charge on the gate of the transistor. The current through the MOSFET will be proportional to the charge on the gate and hence to the intensity of the α particles. If particles of

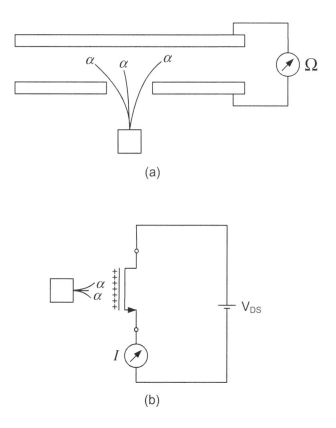

Figure 3.10: (a) Principle of operation of the traditional α-particle smoke detector. (b) Principle of operation of the new detector.

smoke enter the detector and screen the α particles, the positive charge on the gate drops, which leads to a reduced current or total cut-off of the MOSFET.

3.2.1 Structure

The ubiquitous alpha-particle smoke detector has been in wide use for the past four decades. Figure 3.10 (a) shows the basic structure of that popular smoke detector: an α-particle emitter (such as Americium 241) emits α particles into an ionization chamber that consists of two metal electrodes separated by air [10, 29]. The α particles ionize the air molecules, which results in the release of a large number of free electrons inside the chamber. Accordingly, the resistivity of the air inside the chamber decreases, and a small current can be circulated between the two metal electrodes. When smoke particles (which contain a large number of carbon atoms), however, enter the chamber, they quickly attach to the α particles and hence reduce the degree of ionization of the air. The momentary increase in resistivity results in a lower current between the electrodes, which can be detected with the external circuit.

Unfortunately, the change in the resistivity of the air in the traditional smoke detector is quite small in practice, which necessitates the use of a large ionization chamber and an elaborate external circuit to detect the small changes in resistivity that occur when smoke particles enter the detector. Accordingly, the smoke detector has been a traditionally large device. In recent years, however, there has been an interest in a very small smoke detector that can be hidden close to a passenger seat in an aircraft, train, or bus [30]. Such a detector can be useful not only for detecting cigarette smokers, but also for detecting potential terrorists (who have attempted in recent years to use explosive/combustible materials inside public transportation systems). Such a miniature smoke detector was recently introduced by the author. The fundamental principle of the new detector is shown in Figure 3.10 (b). Instead of relying on the principle of ionization of the air, the α particles emitted from the α-particle source are made to strike directly the gate of an n-channel MOSFET transistor. The α particles, being positively charged helium nuclei, will deposit their positive charges on the gate. The MOSFET is known to be highly sensitive to a charge on the gate and will immediately start conducting when the α particles reach the gate. The current I through the MOSFET will be proportional to the charge on the gate, and hence to the number of α particles reaching the gate. If smoke particles enter the detector and get attached to the α particles, a lower positive charge will be present on the gate, which will lead to a lower current or total cut-off of the MOSFET.

Theoretical and experimental investigations will show that this principle is highly sensitive in comparison with the old technique of detecting smoke, and hence this new device can be made substantially smaller than the traditional ionization-based smoke detector (in fact, the new detector is much smaller than the audio buzzer that is typically connected to the detector as an alarming mechanism). It is interesting to point out that optical smoke detectors have been the subject of much research during the past decade [31–35], while radiation-based detectors have not received similar attention (except perhaps for one paper by B. Liu et al. [36] in which the α-particle source was replaced by a beta emitter). Of course, the advantage of radiation-based detectors in comparison with optical detectors is their capability to detect very minute (invisible) amounts of smoke.

3.2.2   Qualitative Description of the Detector

Figure 3.11 shows a schematic of the actual circuit used in the present prototype.

As shown, an α-particle source is arranged such that the emitted α particles strike a small plate that is directly connected to the gate of an n-channel, enhancement mode MOSFET. The α-particle source is a commercially available thin foil containing the ^{241}Am isotope.[2] An optional 10 MΩ resistor can be connected to the gate, as shown, to drain the positive charge from the gate when the α particles are not present. The voltage on the drain terminal of the MOSFET is sensed with a voltage-follower

---

[2] The half-life of ^{241}Am is 432 years.

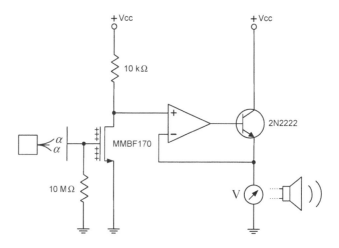

Figure 3.11: Circuit used in the present prototype. A DC buzzer can be connected to the output of the driving transistor to obtain an audible alarm.

circuit consisting of a high-impedance opamp and an NPN transistor. The purpose of the NPN transistor is to act as a load driver (for driving a DC buzzer or other means for indicating the presence of voltage). When the α particles are striking the gate of the MOSFET and a positive charge is present, the MOSFET conducts and hence the output voltage will be low. If, however, the α particles are screened by the presence of smoke and no positive charge is present on the gate, the MOSFET shuts off. But since the input to the opamp is connected to a pull-up resistor, the output voltage in that case will be high (indicating an alarm).

Figure 3.12 shows a photograph of the new detector next to a traditional α-particle smoke detector (for size comparison). The prototype shown is complete except for the optional DC buzzer, which is not included in the photograph.

3.2.3 Theory

Determination of the gate voltage V_{GS}:

The radioactive foil used in the present detector contains approximately 0.1 μCi of the α-particle emitter ^{241}Am (one-tenth of the quantity that is typically used in household smoke detectors). 0.1 μCi is equivalent to 3700 emissions/s. Since the helium nucleus contains two protons, the equivalent of 7400 positive electron charges will be striking the gate of the MOSFET each second. To determine the gate-source voltage V_{GS}, consider the simplified cross-sectional drawing of an n-channel MOSFET shown in Figure 3.13.[3] There are two capacitances of interest in this application. The first is the gate-source capacitance, which can be measured and is usually given in the datasheet of the device. The second capacitance is the gate-channel capaci-

---

[3]Please note: Figure 3.13 is a simplified schematic. Additional details, such as n+ regions, are not shown.

Figure 3.12: Photograph of the new detector (small circuit board on right) next to a traditional α-particle smoke detector.

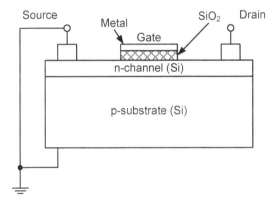

Figure 3.13: Structure of the n-channel MOSFET (simplified).

tance [37], which is a substantially smaller and much harder to determine capacitance (that capacitance must be calculated from an elaborate mathematical model of the device). Due to the bombardment of the gate with α particles and the deposition of positive charges on the gate, electrons will be attracted to the channel by induction. Since there is no well-defined voltage source between the gate and the source terminals, the only capacitance that must be accounted for when attempting to calculate the voltage rise on the gate will be the gate-channel capacitance. The normal approach would be to use the well-known equation $Q = CV$ [38], where Q is the charge on the gate and C is the capacitance, to determine the unknown voltage. Unfortunately, such a calculation cannot be performed since the gate-channel capacitance cannot be determined with reasonable accuracy.

An alternative approach for determining the steady-state gate-source voltage V_{GS} is the following: the steady-state current that will be flowing into the gate terminal will be given by

$$I = \frac{dQ}{dt} = \frac{7400 \times 1.6 \times 10^{-19}}{1} = 1.18 \times 10^{-15} \quad A \tag{3.8}$$

(The optional high-value resistor that is shown in Figure 3.11 is not assumed to be present in this analysis). Under steady-state conditions, this very small current of about 1 fA will constitute a leakage current that will flow to the source terminal, through the insulating silicon dioxide (SiO_2) layer that isolates the gate terminal. The leakage current density J through the SiO_2 layer will be related to the electric field intensity E between the gate and the source terminals by the well-known relationship [38]

$$J = \sigma E \tag{3.9}$$

where σ is the conductivity of SiO_2 (this value is approximately $10^{-16} \ \Omega^{-1} \ m^{-1}$ [39]). The above equation can be written as

$$\frac{I}{A} = \sigma \frac{V_{GS}}{\Delta x} \tag{3.10}$$

where A is the surface area of the gate electrode and Δx is the distance, in general, between the gate and the source (which in practice can be very nonuniform). From Eq. (3.10), V_{GS} can be calculated by using the expression

$$V_{GS} = \frac{I \Delta x}{\sigma A} \tag{3.11}$$

The ratio $\Delta x / A$ can now be determined by knowledge of the gate-source capacitance. According to the datasheet of the MMBF170 transistor used in the present prototype, $C \approx 24$ pF [40]. From the well-known equation [38]

$$C = \varepsilon_0 \varepsilon_r \frac{A}{\Delta x} \tag{3.12}$$

where ε_0 is the permittivity of free space and ε_r is the relative permittivity of the insulating material ($\varepsilon_r = 3.9$ for SiO_2), the ratio $\Delta x / A$ can be calculated to be

$$\frac{\Delta x}{A} = \frac{\varepsilon_0 \varepsilon_r}{C} \approx 1.438 \quad m^{-1} \tag{3.13}$$

By substituting with the quantities determined above into Eq. (3.11), the steady-state V_{GS} is found to be approximately equal to 17 V. Direct measurement of the gate voltage with a Keithley high-impedance electrometer has confirmed this estimated

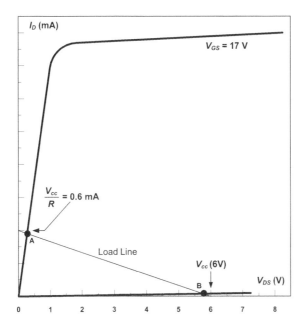

Figure 3.14: Load line for the 10 kΩ pull-up resistor, and the two operating curves of the MOSFET. The drain current I_D is plotted versus the drain-source voltage V_{DS}, with an arbitrary scale for I_D.

voltage.[4] It is to be pointed out that the steady-state gate voltage is typically reached within a fraction of a second in the present application.

Determination of the ON-OFF operating points of the detector:

Figure 3.14 shows the load line for the circuit in Figure 3.11 and the two MOS-FET characteristic curves that are of interest in this application. Because the load attached to the MOSFET (the 10 kΩ pull-up resistor) is large, the MOSFET will be operating in the ohmic region, as Figure 3.14 shows. When $V_{GS} = 0$ (no charge on the gate), the MOSFET will be cut off and $V_{DS} \approx V_{cc}$, which is 6 V in the present prototype. This is the lower operating point in Figure 3.14. To determine the upper operating point (where $V_{GS} = 17$ V), we must solve the following two simultaneous equations:

$$I_D = \frac{V_{cc}}{R} - \frac{1}{R}V_{DS}$$

$$I_D = \frac{V_{DS}}{R_{on}} \tag{3.14}$$

where R is the value of the load resistance (10 kΩ) and R_{on} is the ON resistance of the MOSFET (the inverse of the slope of the characteristic curve shown in

---

[4]The electrometer (Keithley model 6517B) showed a reading of about 8 V, which is to be expected since the impedance of the electrometer is comparable to the impedance of the MOSFET described here.

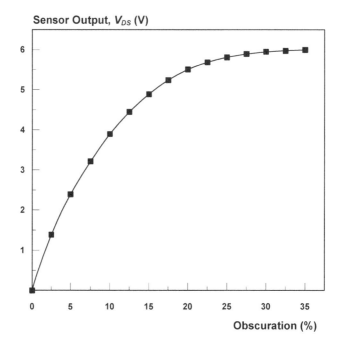

Figure 3.15: MOSFET drain-source voltage (output of the circuit in Figure 3.11) as a function of the obscuration.

Figure 3.14). The first of the above equations is the load-line equation, and the second is the equation representing the characteristic curve. The solution of the two simultaneous equations is

$$V_{DS} = \frac{V_{cc}}{1 + R/R_{on}}$$ (3.15)

with $R = 10$ kΩ, and the MOSFET resistance R_{on} being typically equal to 1 Ω, it is clear that $V_{DS} \approx 0$ when the MOSFET is fully ON.

3.2.4   Experimental Results

Sensor output as a function of obscuration:

   When smoke enters the detector with various densities and screens the α particles, the operating point of the sensor will move from point A in Figure 3.14 to point B. Accordingly, the output voltage (V_{DS}) will increase from 0 to approximately V_{cc} (which is 6 V in the present prototype). Figure 3.15 shows the measured output voltage as a function of the obscuration (in percentage). The level of obscuration was determined at each point in the graph of Figure 3.15 in accordance with standard NFPA-270 [41], where a laser detector[5] was used in a conical radiant source

---

[5]Detector model "Analaser," from Fenwall Protection Systems, Minneapolis, MN.

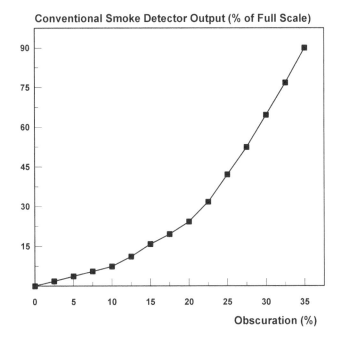

Figure 3.16: Output of a conventional household ionization-based smoke detector (as percentage of full-scale output), as a function of obscuration.

configuration as described in [41] and [42] to measure the smoke obscuration. The results in Figure 3.15 were obtained by testing the sensor in still air. The tests were also conducted by using moving air, and a very small improvement in the sensitivity was observed. Since the improvement is incremental, it will not be reported here.

Figure 3.16 shows the measured output of a conventional household smoke detector (as percentage of full-scale output) as a function of obscuration. By comparing Figures 3.15 and 3.16, it is clear that the new sensor has substantially higher sensitivity at low levels of obscuration. The sensitivity is, however, the same for levels of obscuration higher than 35%.

Sensor output as a function of the distance between the α-particle source and the MOSFET gate:

The output of the sensor was measured as a function of the distance between the α-particle source and the gate of the MOSFET, at a fixed obscuration of 30%. The result is shown in Figure 3.17.

For the present prototype, the distance between the α-particle source and the gate is 5 mm.

Effect of the ambient temperature:

The MOSFET is known to be slightly sensitive to temperature. More specifically, the drain current I_D decreases at temperatures higher than room temperature and

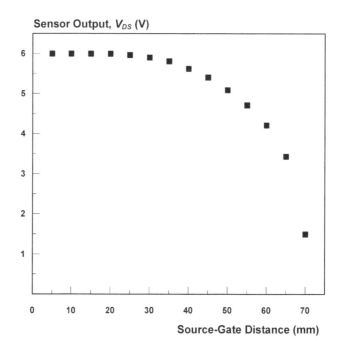

Figure 3.17: Output of the sensor as a function of the distance between the α-particle source and the gate, at a fixed obscuration of 30%.

increases at lower temperatures. For the circuit in Figure 3.11, the result is that the output voltage V_{DS} drifts higher and lower with temperature. To test the effect of temperature variation, the sensor was placed inside a variable temperature chamber and the tests shown in Figure 3.15 were repeated. Figure 3.18 shows the results for temperatures of $+80°C$ and $-40°C$. Clearly, the effect of the ambient temperature on the performance of the sensor is minimal.

3.2.5   Conclusion

The new α-particle smoke detector described in this section is substantially smaller and substantially more sensitive than conventional smoke detectors that depend on the principle of air ionization. The principle used in the new detector, namely, the transfer of the charge of the α particles to the gate of a MOSFET, allows the construction of a detector that is characterized by small size and high sensitivity in comparison with the ionization chamber that is used in conventional α-particle-based detectors. As Figure 3.15 shows, the sensitivity of the new detector is substantially high at low levels of smoke obscuration. The sensor will be very useful for detecting smoke in tight places, such as next to a passenger's seat in a bus or airplane. As indicated earlier in this section, the advantage of radiation-based detectors in comparison with optical detectors is their capability of detecting very minute amounts of smoke.

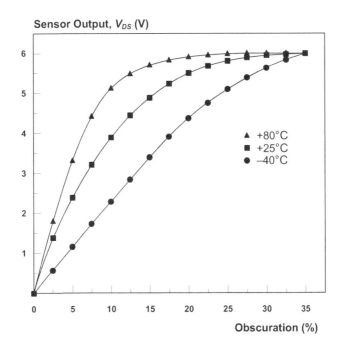

Figure 3.18: MOSFET drain-source voltage (output of the circuit in Figure 3.11) as a function of the obscuration, at temperatures of +80°C, +25°C, and -40°C.

3.3 Quiz

1) A miniature carbon monoxide (CO) gas sensor that works at room temperature is based on which technological principle?
Ans: Reaction of CO gas with gold nanoparticles.

2) True or false: The presence of CO gas is detected in the room-temperature miniature CO gas sensor as an increase in the resistivity of a carbon nanotube array.
Ans: False. The effect is detected as a decrease in the resistivity of a carbon nanotube array.

3) A novel smoke detector that uses an alpha-particle source is based on which scientific principle?
Ans: Replacement of the traditional ionization chamber with a CMOS transistor.

References

[1] United States Occupational Safety and Health Administration, *Occupational Safety and Health Guideline for Carbon Monoxide*, http://www.osha.gov/SLTC/healthguidelines/carbonmonoxide/recognition.html, 2010.

[2] G.L. Anderson and D.M. Hadden, *The Gas Monitoring Handbook*, Avocet Press, New York, 1999.

[3] S.C. Hung, C.W. Chen, C.Y. Shieh, G.C. Chi, R. Fan, and S.J. Pearton, High sensitivity carbon monoxide sensors made by zinc oxide modified gated GaN/AlGaN high electron mobility transistors under room temperature, *Applied Physics Letters*, vol. 98, art. no. 223504, 2011.

[4] N. Lopez and J.K. Norskov, Catalytic CO oxidation by a gold nanoparticle: A density functional study, *J. of the American Chemical Society*, vol. 124, pp. 11262–11263, 2002.

[5] M. Haruta and M. Date, Advances in the catalysis of Au nanoparticles, *Applied Catalysis A*, vol. 222, pp. 427–437, 2001.

[6] H.C. Choi, M. Shim, S. Bangsaruntip, and H. Dai, Spontaneous reduction of metal ions on the sidewalls of carbon nanotubes, *J. of the American Chemical Society*, vol. 124, pp. 9058–9059, 2002.

[7] E. Becker, M. Andersson, M. Eriksson, A.L. Spetz, and M. Skoglundh, Study of the sensing mechanism towards carbon monoxide of platinum-based field effect sensors, *IEEE Sensors Journal*, vol. 11, 7, pp. 1527–1534, 2011.

[8] Feng-Renn Juang, Yean-Kuen Fang, Yen-Ting Chiang, Tse-Heng Chou, Cheng-I Lin, Cheng-Wei Lin, and Yan-Wei Liou, Comparative study of carbon monoxide gas sensing mechanism for the LTPS MOS schottky diodes with various metal oxides, *IEEE Sensors Journal*, vol. 11, 5, pp. 1227–1232, 2011.

[9] S. Bicelli, A. Depari, G. Faglia, A. Flammini, A. Fort, M. Mugnaini, A. Ponzoni, V. Vignoli, and S. Rocchi, Model and experimental characterization of the dynamic behavior of low-power carbon monoxide MOX sensors operated with pulsed temperature profiles, *IEEE Transactions on Instrumentation and Measurement*, vol. 58, 5, pp. 1324–1332, 2009.

[10] E. Cordos, L. Ferenczi, S. Cadar, S. Costiug, G. Pitl, A. Aciu, A. Ghita, and M. Chintoanu, Methane and carbon monoxide gas detection system based on semiconductor sensor, *IEEE Int. Conf. on Automation, Quality and Testing, Robotics*, pp. 208–211, 2006.

[11] J. Tan, W. Wlodarski, K. Kalantar-Zadeh, and P. Livingston, Carbon monoxide gas sensor based on titanium dioxide nanocrystalline with a langasite substrate, *IEEE Sensors Conf.*, pp. 228–231, 2006.

[12] G.G. Mandayo, E. Castano, and F.J. Gracia, Carbon monoxide detector fabricated on the basis of a tin oxide novel doping method, *IEEE Sensors Journal*, vol. 2, 4, pp. 322–328, 2002.

[13] V. Blondeau-Patissier, M. Vanotti, D. Rabus, S. Ballandras, M. Chkounda, J.M. Barbe, and J.Y. Rauch, Development of accurate system of gas detection based

on Love wave sensors functionalized with cobalt corroles applied to the detection of carbon monoxide, *IEEE Sensors Conf.*, pp. 1078–1081, 2011.

[14] Sang-youn Park and Dong-wook Yoon, A study on the adsorption of carbon monoxide on silica supported Ru-Fe alloy, *Proceedings of the 7th Korea-Russia International Symposium on Science and Technology*, vol. 3, pp. 228–238, 2003.

[15] M.L. Homer, A.V. Shevade, H. Zhou, A.K. Kisor, L.M. Lara, S.Yen, and M.A. Ryan, Polymer-based carbon monoxide sensors, *IEEE Sensors Conf.*, pp. 1504–1508, 2010.

[16] Ma. Cheng, B.L. Scott, G.R. Pickrell, and Anbo Wang, Porous capillary tubing waveguide for high-temperature carbon monoxide detection, *IEEE Photonics Technology Letters*, vol. 22, 5, pp. 323–325, 2010.

[17] Yu-Kai He, Ru-Lin Wang, and Yu-Qiang Yang, A novel carbon monoxide detection system based on infrared absorption used in mine, *IEEE Int. Conf. on Machine Learning and Cybernetics*, pp. 645–649, 2006.

[18] Shanying Zhu, Youping Chen, Gang Zhang, and Jiming Sa, An optical fiber sensor based on absorption spectroscopy for carbon monoxide detection, *IEEE Int. Conf. on Computer Design and Applications (ICCDA)*, vol. 2, pp. 589–592, 2010.

[19] Jingchao Zhang, Suxia Cao, Lijun Guan, Carbon monoxide gas sensor based on cavity enhanced absorption spectroscopy and harmonic detection, *IEEE Symposium on Photonics and Optoelectronics (SOPO)*, pp. 1–4, 2009.

[20] J. Chen, A. Hangauer, R. Strzoda, M. Ortsiefer, M. Fleischer, and M.C. Amann, Compact carbon monoxide sensor using a continuously tunable 2.3 m single-mode VCSEL, *21st Annual Meeting of the IEEE Lasers and Electro-Optics Society (LEOS)*, pp. 721–722, 2008.

[21] N.A. Salmon and R. Appleby, Carbon monoxide detection using passive and active millimeter wave radiometry, *IEEE Int. Conf. on Microwave and Millimeter Wave Technology (ICMMT)*, pp. 522–525, 2000.

[22] S.M. Walas, *Reaction Kinetics for Chemical Engineers*, McGraw Hill, New York, 1959.

[23] M.J. Pilling and P.W. Seakins, *Reaction Kinetics*, Oxford University Press, New York, 1995.

[24] J. Szanyi, W.K. Kuhn, and D.W. Goodman, CO oxidation on palladium: A combined kinetic-infrared reflection absorption spectroscopic study of Pd(111), *J. Phys. Chem*, vol. 98, 11, pp. 2978–2981, 1994.

[25] V. Iablokov, K. Frey, O. Geszti, and N. Kruse, High catalytic activity in CO oxidation over MnOx nanocrystals, *Catal. Lett.*, vol. 134, pp. 210–216, 2010.

[26] M. Haruta, S. Tsubota, T. Kobayashi, H. Kageyama, M.J. Genet, and B. Delmon, Low-temperature oxidation of CO over gold supported on TiO_2, Fe_2O_3, and Co_3O_4, *J. of Catalysis*, vol. 144, 1, pp. 175–192, 1993.

[27] M. Haruta, Size and support dependency in the catalysis of gold, *Catalysis Today*, vol. 36, 1, pp. 153–166, 1997.

[28] S.A. Dyer, *Survey of Instrumentation and Measurement*, Wiley, New York, 2001.

[29] W. Boyes, Instrumentation reference book, *Butterworh-Heinemann/Elsevier*, Burlington, MA, 2010.

[30] C. Hipsher and D. Ferguson, *Fire Protection*, Boeing Corporation *AERO* Magazine, Qtr. 2, pp. 11–19, 2011.

[31] Jimin Cheon, Jeonghwan Lee, Inhee Lee, Youngcheol Chae, Youngsin Yoo, and Gunhee Han, A single-chip CMOS smoke and temperature sensor for an intelligent fire detector, *IEEE Sensors Journal*, vol. 9, 8, pp. 914–921, 2009.

[32] Z.J. Aleksic, The analysis of the transmission-type optical smoke detector threshold sensitivity to the high rate temperature variations, *IEEE Trans. on Instrumentation and Measurement*, vol. 53, 1, pp. 80–85, 2004.

[33] E.D. Lester and A. Ponce, An anthrax "smoke" detector, *IEEE Engineering in Medicine and Biology Magazine*, vol. 21, 5, pp. 38–42, 2002.

[34] Z.J. Aleksic, Evaluation of the design requirements for the electrical part of transmission-type optical smoke detector to improve its threshold stability to slowly varying influences, *IEEE Trans. on Instrumentation and Measurement*, vol. 49, 5, pp. 1057–1062, 2000.

[35] Z.J. Aleksic, Minimization of the optical smoke detector false alarm probability by optimizing its frequency characteristic, *IEEE Trans. on Instrumentation and Measurement*, vol. 49, 1, pp. 37–42, 2000.

[36] Baojun Liu, D. Alvarez-Ossa, N.P. Kherani, S. Zukotynski, and K.P. Chen, Gamma-free smoke and particle detector using tritiated foils, *IEEE Sensors Journal*, vol. 7, 6, pp. 917–918, 2007.

[37] D. Neamen, *An Introduction to Semiconductor Devices*, McGraw Hill, New York, 2007.

[38] W.H. Hayt and J.A. Buck, *Engineering Electromagnetics*, McGraw Hill, New York, 2006.

[39] James F. Shackelford and William Alexander, *CRC Materials Science and Engineering Handbook*, CRC Press, Boca Raton, FL, 2000.

[40] Fairchild Semiconductor, BS170/MMBF170 N-channel enhancement mode field effect transistor, *Fairchild Semiconductor datasheet*, 2010, www.fairchildsemi.com.

[41] National Fire Protection Association, NFPA 270—Standard test method for measurement of smoke obscuration using a conical radiant source in a single closed chamber, *NFPA*, 2002, www.nfpa.org.

[42] Paul E. Patty, A scientific approach to characterize smoke from flaming and smoldering fires, *Underwriters Laboratories Inc.*, 2010, www.ul.com.

Chapter 4

Moisture Sensors

Moisture sensors exist in a number of varieties: capacitive [1–7], resistive [6], microwave based [7, 13, 14, 16], in addition to other more sophisticated varieties [8–10, 12, 14–16, 24]. Among the technologies used at present, the variable-capacitance-based moisture sensor is particularly noteworthy because of its low cost and the simplicity of its design. A drawback of this type of sensor, however, is that the variable capacitor used must be considerably large in order to obtain sufficient sensitivity to the presence of moisture [1–7]. This limits the range of applications for this type of sensor. This chapter introduces a new ultraminiature, ultrahigh sensitivity moisture sensor that is based on ultracapacitor technology (a direct nanotechnology application). An ultracapacitor is assembled from ordinary activated carbon electrodes; however, unlike ordinary ultracapacitors, no liquid electrolyte exists inside the device. The ultracapacitor electrodes are placed on both sides of a layer of porous silicon in which KOH (potassium hydroxide) in powder form is embedded. As moisture penetrates the porous silicon layer, a liquid electrolyte starts to form. The conductivity of the electrolyte, and hence the capacitance of the ultracapacitor assembly, increase as the amount of moisture increases. In the present prototype, an increase in the relative humidity from 5% to 80% results in a capacitance variation from 0 to 17 μF. The sensor can be very useful for monitoring moisture penetration inside small electronic devices and packages that are sensitive to moisture.

4.1   Structure

The new moisture sensor is based on the idea of using an ultracapacitor structure rather than an ordinary capacitor. Inside the ultracapacitor, the moisture itself constitutes the electrolyte that is needed for charge transfer. This concept is illustrated in Figure 4.1.

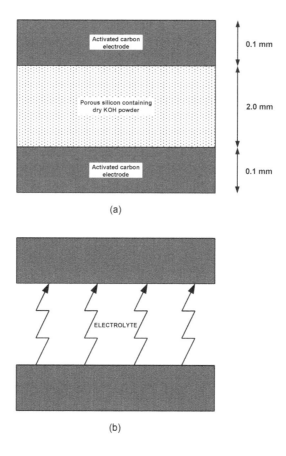

Figure 4.1: Fundamental concept of the new moisture sensor. (a) An ultracapacitor is assembled by using two activated carbon electrodes and an insulating layer of porous silicon. Inside the pores of the silicon layer, KOH (potassium hydroxide) in powder form is embedded. When the assembly is dry, no electrolyte exists inside the ultracapacitor and hence the capacitance of the device is equal to zero. (b) In the presence of moisture, the combination of KOH and water becomes an aqueous electrolyte for the ultracapacitor. The conductivity of the electrolyte and hence the capacitance of the device increase as the moisture content increases. In the present prototype, the capacitance varies from 0 to approximately 17 μF as the relative humidity is increased from 5% to 80%.

It is to be pointed out that the sensor just described is not a fast-response sensor. The fastest sensors for moisture detection remain those that are based on microwave techniques [7, 13, 14, 16]. This new sensor, however, has the highest sensitivity with respect to size among all known types of moisture sensors (including capacitive, microwave, and other sensors). This sensor will therefore be very useful in applications such as the detection of moisture in small electronic packages and components,

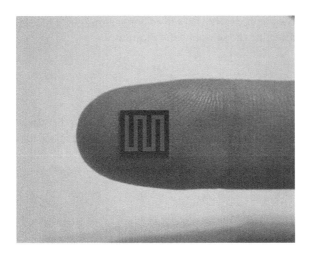

Figure 4.2: Photograph of the sensor, resting on the author's finger. The sensor measures 8 mm × 8 mm.

detection of moisture inside composite aircraft materials, and other embedded moisture sensing applications where high sensitivity and a miniature size are desirable.

Figure 4.2 shows a photograph of the sensor (one of several prototypes fabricated by the author). The prototype shown in the figure measures 8 mm × 8 mm and its total thickness is approximately 2 mm. The sensor can be made much smaller. A 2 mm × 2 mm sensor exhibits a total capacitance variation of about 1 μF (still a substantially large and useful capacitance variation).

As shown in the photograph, the activated carbon electrodes do not occupy the entire surface area of the porous silicon slab, in order to allow moisture to seep through the pores of the slab.[1] The activated carbon electrodes were deposited with screen printing, and the thickness of each electrode is approximately 0.1 mm. The electrodes deposited with this technique have an internal surface area of approximately 500 m^2 per gram of the material [17, 25].

The porous silicon slab was obtained from a commercial supplier of porous silicon. The mean pore size (diameter) of the pores in the slab is approximately 10 nm. To embed the KOH powder inside the slab, KOH at a concentration of 35% by weight was first dissolved in distilled water, and the slab was soaked in the solution. The slab was then sintered in a furnace at a temperature of 500°C for 30 min to remove the water. The activated carbon electrodes were subsequently deposited on both sides of the slab.

---

[1] The pattern shown on the sensor is applied to both sides. Testing has shown that the moisture penetration into the sensor is essentially unaffected by the presence of the pattern (or carbon electrode), due to the very complex "web" of pores inside the porous silicon. Moisture easily penetrates inside the porous silicon layer even if the available surface area is small.

Sensors thus assembled were kept in an argon atmosphere to prevent the ambient moisture from affecting the sensors before testing.

4.2 Theory

The objective now is to obtain a mathematical relationship between the capacitance of the ultracapacitor and the conductivity of the electrolyte inside the device. We first point out that the current density J in any medium is related to the conductivity σ by the following well-known equation [24]

$$J = \sigma E \qquad (4.1)$$

where E is the electric field intensity. J and E are further related to the current I and the voltage V by the relations

$$J = \frac{I}{A} \quad \text{and} \quad E = \frac{V}{d} \qquad (4.2)$$

where A in the present application is the cross-sectional area of the porous slab and d is its thickness. Hence,

$$\frac{I}{A} = \sigma \frac{V}{d} \qquad (4.3)$$

The current through the electrolyte will be related to the rate of flow of charges by

$$I = \frac{dQ}{dt} \qquad (4.4)$$

where Q is the ionic charge moving through the cross-sectional area of the slab. From the previous two equations, we have

$$\frac{dQ}{dt} = \frac{AV}{d} \, \sigma(t) \qquad (4.5)$$

In the above equation, the conductivity is assumed to be a time-varying quantity (the conductivity will be indeed slowly varying as moisture seeps through the device), and the voltage V across the terminals of the ultracapacitor is assumed to be constant. Hence,

$$Q = \frac{AV}{d} \int \sigma(t) dt \qquad (4.6)$$

The accumulated charge Q at each of the two interfaces in the ultracapacitor is related to the capacitance of the interface by the well-known relationship [24]

$$Q = C_{int} V_{int} \qquad (4.7)$$

where C_{int} is the capacitance of the interface and V_{int} is the voltage across the interface. We now point out that $V_{int} = V/2$, where V is the total voltage applied to the device. From Eqs. (4.6) and (4.7) we now have

$$\frac{1}{2} C_{int} V = \frac{AV}{d} \int \sigma(t) dt \qquad (4.8)$$

or

$$C_{int} = \frac{2A}{d} \int \sigma(t)dt \qquad (4.9)$$

But the total capacitance C is equal to $C_{int}/2$ (since the configuration is essentially that of two capacitors in series) [25]. Accordingly,

$$C = \frac{A}{d} \int \sigma(t)dt \qquad (4.10)$$

It is to be pointed out that the integral of σ over time is dimensionally equivalent to the permittivity (ε) of a medium. In the present case, it is extremely difficult to determine the capacitance from the well-known equation $C = \varepsilon A/d$, since the capacitance of an ultracapacitor is actually a function of the internal surface area of the activated carbon electrodes.[2] While the internal surface area can actually be estimated with good accuracy, it is only partially utilized in the present device (as charges accumulate over time, the internal surface area eventually becomes fully utilized). Since the estimation of the internal surface area that is effectively utilized is very difficult, Eq. (4.10) should be used for the calculation of the capacitance of the device rather than the traditional capacitance equation. By knowledge of the conductivity as a function of time, the capacitance can be determined at any point in time.[3] Conversely, as shown in the following section, the capacitance can be measured and correlated to the total amount of moisture that has penetrated the device.

4.3 Main Experimental Results

The sensor was tested in a computer-controlled chamber that features variable humidity and variable temperature.[4] Figure 4.3 shows the measured conductivity σ of the porous silicon slab at 25°C, as the relative humidity was increased from 5% to 80%. Figure 4.4 shows the same measurements, repeated at temperatures of 10°C and 70°C. It should be pointed out that the fact that water solidifies at 0°C and boils at 100°C, in combination with the fact that most humidity control chambers can reliably control the humidity in the range of 10 to 70°C, make this particular range of temperatures (10–70°C) the most suitable range for testing a moisture sensor.

Figure 4.5 shows the measured capacitance C of the sensor at 25°C, as the relative humidity was increased from 5% to 80%. Figure 4.6 shows the same measurements, repeated at temperatures of 10°C and 70°C.

In order to understand and visualize how the results shown above correlate with Eq. (4.10), plots of the variation of the conductivity and capacitance as functions of time are shown in Figure 4.7, as the relative humidity is increased as a step-function from 5% to 10%.

---

[2]Furthermore, the distance d that would be utilized in the traditional equation is the distance between the opposite species of charges, which is approximately 1 nm.

[3]In comparison with other capacitive moisture sensors [6, 7], it should be clear that, in the present sensor, an increasing conductivity leads to an increase in capacitance.

[4]Chamber model ESX-3CA, from ESPEC Inc.

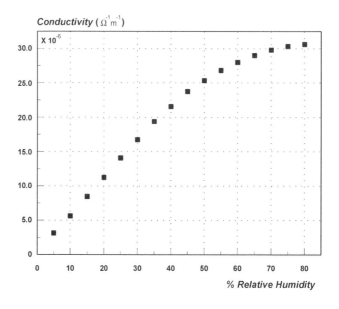

Figure 4.3: Conductivity of the porous silicon slab at 25°C, as the relative humidity is increased from 5% to 80%.

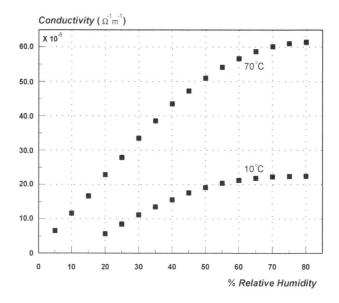

Figure 4.4: Conductivity of the porous silicon slab at 10°C and 70°C, as the relative humidity is increased from 5% to 80%.

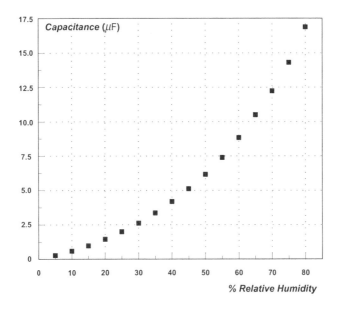

Figure 4.5: Capacitance of the sensor at 25°C, as the relative humidity is increased from 5% to 80%.

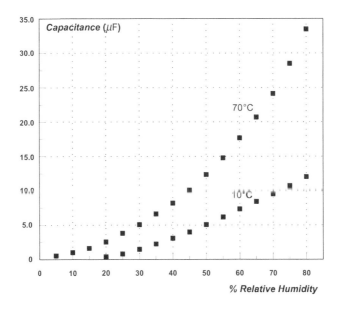

Figure 4.6: Capacitance of the sensor at 10°C and 70°C, as the relative humidity is increased from 5% to 80%.

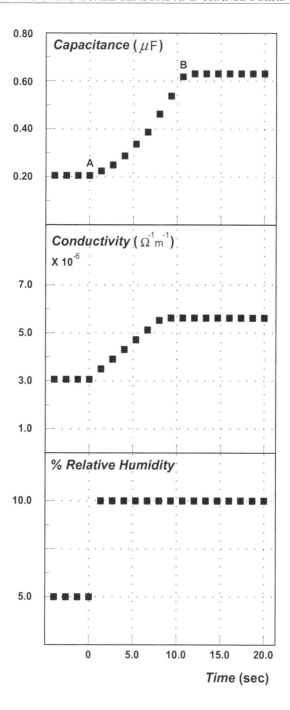

Figure 4.7: Change in conductivity and capacitance in response to a unit-step rise in relative humidity from 5% to 10%. The testing temperature was 25°C.

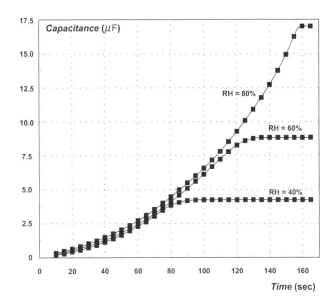

Figure 4.8: Variation of the capacitance C as a function of time, at a temperature of 25°C and for three fixed values of relative humidity.

As the plots show, σ increases linearly from approximately 3×10^{-6} $\Omega^{-1}m^{-1}$ to 5.6×10^{-6} $\Omega^{-1}m^{-1}$, whereafter it remains constant. The corresponding capacitance increase between points A and B on the upper plot was found to be in perfect agreement with Eq. (4.10) (similar step-function tests were conducted for other ranges of the relative humidity, and the results were always in perfect agreement with Eq. (4.10)).

The result in Figure 4.7 does, however, deviate from Eq. (4.10) beyond point B on the plot. Whereas Eq. (4.10) predicts that the capacitance will keep increasing, the capacitance actually stabilizes at a constant value of about 0.63 μF, as the plot shows. The reason for this deviation from the expected behavior is the following: the porous silicon slab is a complicated medium. Due to the structure of the pores, certain paths within the slab become conductive as the relative humidity is increased, while other paths do not accumulate sufficient moisture to become conductive. Once all the available charges within a conductive path migrate to the activated carbon electrodes, the path becomes "saturated" and no further migration of charges can occur. The capacitance therefore remains constant. As Figures 4.5 and 4.6 show, the capacitance does indeed increase once again as the relative humidity is further increased to higher levels.

Figure 4.8 shows the variation of the capacitance of the sensor as a function of time, for different relative humidity values and at a fixed temperature of 25°C.

Figure 4.9 shows temperature hysteresis curves that were obtained at two fixed values of the relative humidity (80% and 20%). The figures show the capacitance that was measured as the temperature was cycled. The maximum hysteresis error

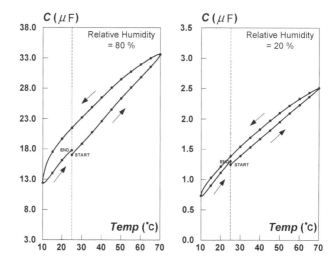

Figure 4.9: Temperature hysteresis curves. The maximum hysteresis error in the measured capacitance of the sensor is approximately 12% FSO, as the temperature is cycled from 10°C to 70°C.

is approximately 12% of the full-scale output, as the figures show. Given that this error is actually less than the typical tolerance error of most capacitors, it is clear that the measured capacitance of the sensor will be a reliable indicator of the amount of moisture that penetrates the device, regardless of the temperature cycling conditions.

Figure 4.10 shows a block diagram of the interface circuit that was used in the various tests conducted. The circuit is a well-known 555 oscillator circuit that converts an unknown capacitance to frequency (the circuit is described in references such as [19]). It is to be pointed out that the interface circuit does not amplify or compensate for the characteristics of the sensor.

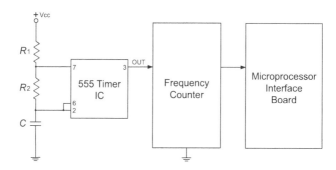

Figure 4.10: Block diagram of the interface circuit used to measure the capacitance C.

Figure 4.11: Photograph of the sensor embedded inside an open-cavity IC chip. In a second IC chip the cavity is shown after it was filled with the epoxy material.

4.4 Auxiliary Experimental Results

Testing of the sensor inside an encapsulated IC chip:

To test the performance of the sensor in a real application, the sensor was embedded inside an open-cavity IC chip. The cavity measures 8 mm × 8 mm (same as the dimensions of the prototype shown in Figure 4.2). Figure 4.11 shows the IC chip and the embedded sensor. The cavity was subsequently filled with an ultraviolet-curable epoxy.[5] The epoxy material used is very similar to the ones studied by Galloway and Miles. [26] and by Fan et al. [27]. It has a diffusivity constant of approximately 10^{-4} cm^2/s at 25°C and a concentration ratio constant of approximately 2×10^{-2}. The purpose of this test was to compare the performance of the sensor to the predictions that can be made based on the law of diffusion [26–28].

Very detailed predictions for the diffusion of moisture inside porous ceramic and polymer materials were reported in references such as [26], [27], and [28]. The purpose of this section is not to repeat those studies, but to demonstrate the performance of the sensor in view of the predictions made by those studies. The IC chips shown in Figure 4.11 were exposed to an ultraviolet light source, and, after the epoxy solidified, they were exposed to precisely controlled humidity at 25°C. Figure 4.12 shows the results of two weight measurement tests that were conducted at 80% RH and at 40% RH. The percentage weight gain shown in the figure over a 7-day period is the weight gain of the epoxy material, excluding the rest of the IC chip. The percentage weight gain was determined by weighing the IC chip every 12 hours on a balance

---

[5] Available from Dymax Corp., Torrington, CT.

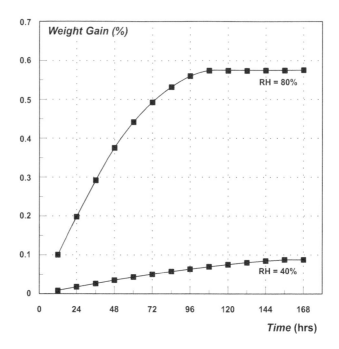

Figure 4.12: Percentage weight gain of the epoxy material at 25° C, for ambient relative humidities of 40% and 80%.

with an accuracy of 0.01 mg. The results shown in Figure 4.12 are very close to those reported by Galloway and Miles. for a similar type of epoxy material [26].

Figure 4.13 shows the values of the measured capacitance of the sensor for the tests shown in Figure 4.12. As can be observed, the results are within ± 5% of the results shown in Figure 4.8, even though the time scale is substantially different. These results demonstrate that the measurements provided by the sensor are in accord with predictive moisture penetration calculations that can be made for different materials based on the law of diffusion. More specifically, as can be concluded collectively from Figures 4.12 and 4.13, the capacitance reading of the sensor can be related to an equivalent ambient moisture concentration, which in turn can be related to percentage weight gain in the material in which the sensor is embedded. The percentage weight gain can then be directly related to the concentration of water inside the material by equations such as those given in [26] and [27].

Relative humidity hysteresis testing:

A relative humidity hysteresis test was conducted by cycling the relative humidity at constant temperature. It was found that once the capacitance of the sensor reaches a peak value, it does not decrease quickly when the relative humidity is subsequently lowered. Rather, it can take a period of several hours for the capacitance to settle to a lower value that corresponds to lower relative humidity. This is due to the fact

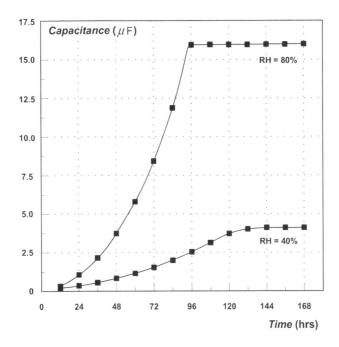

Figure 4.13: Measured capacitance of the sensor as a function of time, for the tests shown in Figure 4.12.

that KOH has an excellent capability to retain moisture. Accordingly, this sensor will be useful as an indicator of the penetration of moisture in certain devices and applications, but will not be useful in other applications that require quick, real-time monitoring of the relative humidity.

Effect of contaminants (organic vapors and liquids) on the performance of the sensor:

To test the effect of organic contaminants on the performance of the sensor, a system was devised in which the sensor was placed in a small vacuum chamber that is connected to a computer-controlled evaporator. The evaporator, in turn, is connected to a cylinder that supplies dry N_2 gas. The N_2 gas is circulated through the chamber and mixed with the vapors emerging from the evaporator. The control system maintains precise control of the concentration of the contaminant vapor within the N_2 gas (specified in parts-per-million, or ppm). Three solvents were tested: benzene, ethanol, and water. Figure 4.14 shows the results, where the measured capacitance of the sensor is plotted against the concentration of the vapor in ppm. In the figure, the concentrations shown correspond with the relative humidity values shown in Figure 4.5; that is, 3100 ppm = 10% RH, 6200 ppm = 20% RH, etc.

The results shown in Figure 4.14 are to be expected, since KOH has lower solubility in ethanol in comparison with water and has zero solubility in benzene (in fact,

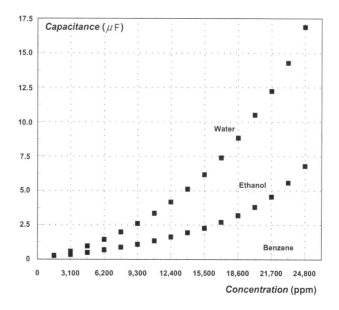

Figure 4.14: Capacitance versus concentration for three different solvents.

the capacitance measured was exactly 0 in the presence of benzene). The conclusion therefore is that the presence of organic contaminants will degrade the performance of the sensor. Of course, the degradation will depend on the ratio of the contaminant to the water present in the moisture.

Testing of the long-term aging effects:

Aging was simulated by subjecting the sensor to 1000 temperature cycles (such as the cycle shown in Figure 4.9). The tests, which were conducted over a period of nearly 60 days, were repeated for different levels of relative humidity. The results (i.e., the hysteresis in the measured capacitance) did not vary significantly from the results shown in Figure 4.9.

4.5 Conclusion

A novel miniature moisture sensor was presented in this chapter. This new sensor is now in production and should be available commercially in early 2015. Based on ultracapacitor technology, the sensor has the highest sensitivity with respect to size among all known types of moisture sensors [20, 21]. The prototype device described here has exhibited a change in capacitance of approximately 17 μF over the 5%–80% relative humidity range. The capacitance measurements have proved to be quite repeatable and reliable. This new sensor will be very useful in applications such as the detection of moisture inside small electronic packages and components, the detection of moisture inside composite aircraft materials, as well as other embedded moisture sensing applications where high sensitivity and a miniature size are desirable.

4.6   Quiz

1) True or false: Highly sensitive, miniature moisture sensors detect the variation in capacitance between two electrodes.
Ans: False. This new type of moisture sensor detects the variation of capacitance of an ultracapacitor as moisture penetrates the device and changes the properties of the electrolyte.

2) True or false: The new type of moisture sensor that detects the variation of an ultracapacitor is very fast and is also the smallest and the most sensitive among all the known types of moisture sensors.
Ans: False. This new type of sensor is the smallest and the most sensitive, but it is a slow-acting sensor in comparison with other types of moisture sensors.

References

[1] Robert B. McIntosh and Mark E. Casada, Fringing field capacitance sensor for measuring the moisture content of agricultural commodities, *IEEE Sensors J.*, vol. 8, 3, pp. 240–247, 2008.

[2] Chari V. Kandala and Jaya Sundaram, Nondestructive measurement of moisture content using a parallel-plate capacitance sensor for grain and nuts, *IEEE Sensors J.*, vol. 10, 7, pp. 1282–1287, 2010.

[3] Willy Ludurczak, Claude Pellet, Olivier Garel, Elisabeth Dufour-Gergam, and Fabrice Verjus, Influence on moisture sensor performances, and characterization of different specific area porous silicon layers, *IEEE Sensors 2007 Conference*, p. 1245-1250.

[4] S. Dhanekar, P.M.Z. Hasan, S. Hussain, T. Islam, S.S. Islam, K. Sengupta, and D. Saha, Study of cross-sensitivity of porous alumina based trace moisture sensor in dry gases, *IEEE 3rd International Conference on Sensing Technology, Tainan, Taiwan*, p. 656-660, 2008.

[5] D. Wobschall and D. Lakshmanan, Wireless soil moisture sensor based on fringing capacitance, *IEEE Sensors 2005 Conference*, p. 1-4.

[6] David R. Day and David D. Shepard, A microdielectric analysis of moisture diffusion in thin epoxy/amine films of varying cure state and mix ratio, *Polymer Engineering and Science*, vol. 32, 8, pp. 524–528, 1992.

[7] David R. Day and David D. Shepard, RP034—Moisture diffusion monitoring in polymers with microdielectric sensors, *Micromet Instruments, Inc.*, 2011.

[8] Ben-Je Lwo and Chih-Shiang Lin, Measurement of moisture-induced packaging stress with piezoresistive sensors, *IEEE Trans. on Advanced Packaging*, vol. 30, 3, pp. 393–401.

[9] Mohamed M. Ghretli, Kaida Khalid, Ionel V. Grozescu, M. Hamami Sahri, and Zulkifly Abbas, Dual-frequency microwave moisture sensor based on circular microstrip antenna, *IEEE Sensors J.*, vol. 7, 12, pp. 1749–1756, 2007.

[10] Samir Trabelsi, Stuart 0. Nelson, and Omar Ramahi, A low-cost microwave moisture sensor, *Proceedings of the 36th European Microwave Conference, Manchester*, UK, pp. 447–450, 2006.

[11] F. Jafari, K. Khalid, W.M. Daud, M. Yusoff, and J. Hassan, Development and design of microstrip moisture sensor for rice grain, *IEEE RF and Microwave Conference*, pp. 258–262, 2006.

[12] Kaida Khalid, Mohamed M. Ghretli, Zulkifly Abbas, and Ionel Grozescu, Development of planar microwave moisture sensors for hevea rubber latex and oil palm fruits, *IEEE RF and Microwave Conference*, pp. 10–14, 2006.

[13] Huichun Xing, Jing Li, R. Liu, E. Oshinski, and R. Rogers, 2.4 GHz On-board parallel plate soil moisture sensor system, *IEEE Sensors for Industry Conference*, pp. 35–39, 2005.

[14] Jiuqing Liu, Resonator sensor for moisture content measurement, *The Sixth World Congress on Intelligent Control and Automation*, pp. 4982–4986, 2006.

[15] Tomas Unander and Hans-Erik Nilsson, Characterization of printed moisture sensors in packaging surveillance applications, *IEEE Sensors J.*, vol. 9, 8, pp. 922–928, 2009.

[16] J.H. Rodriguez-Rodriguez, F. Martinez-Pinon, J.A. Alvarez-Chavez, and D. Jaramillo-Vigueras, Polymer optical fiber moisture sensor based on evanescent-wave scattering to measure humidity in oil-paper insulation in electrical apparatus, *IEEE Sensors 2008 Conference*, pp. 1052–1056.

[17] H.E. Nilsson, J. Siden, T. Unander, T. Olsson, P. Jonsson, A. Koptioug, and M. Gulliksson, Characterization of moisture sensor based on printed Carbon-zinc energy cell, *IEEE 5th International Conference on Polymers and Adhesives in Microelectronics and Photonics, Polytronic*, pp. 82–86, 2005.

[18] Qi Chen, Ji Fang, Hai-Feng Ji, and K. Varahramyan, Microfabrication and characterization of SiO_2 microcantilever for high sensitive moisture sensor, *IEEE Sensors 2007 Conference*, pp. 1436–1440.

[19] M. Morisawa and H. Yokomori, Improvement in the sensitivity of dye-doped POF-type moisture sensor, *IEEE Sensors 2010 Conference*, pp. 508–512.

[20] N.K. Pandey, K. Tiwari, and A. Roy, Cu_2O doped ZnO as moisture sensor, *IEEE Sensors 2009 Conference*, pp. 312–316.

[21] K. Katoh, Y. Ichikawa, E. Iwase, K. Matsumoto, and I. Shimoyama, Thermal-based skin moisture device with contact pressure sensor, *IEEE 23rd International Conference on Micro Electro Mechanical Systems (MEMS)*, pp. 276–280, 2010.

[22] B.E. Conway, Electrochemical supercapacitors, Kluwer Academic Publishers, New York, 1999.

[23] A. Burke, Ultracapacitors: why, how, and where is the technology, *J. of Power Sources*, vol. 91, pp. 37–50, 2000.

[24] W.H. Hayt and J.A. Buck, *Engineering Electromagnetics*, McGraw Hill, New York, 2006.

[25] E.A. Parr, *IC 555 Projects*, Babani Publishing, Ltd., London, UK, p. 13, 1981.

[26] J.E. Galloway and B.M. Miles, Moisture absorption and desorption predictions for plastic ball grid array packages, *IEEE Trans. on Components, Packaging, and Manufacturing Technology—Part A*, vol. 20, 3, pp. 274–279, 1997.

[27] Xuejun Fan, G.Q. Zhang, W.V. Driel, and L.J. Ernst, Analytical solution for moisture-induced interface delamination in electronic packaging, *IEEE Electronic Components and Technology Conference*, pp. 733–738, 2003.

[28] G.S. Ganesan and H.M. Berg, Model and analyses for solder reflow cracking phenomenon in SMT plastic packages, *IEEE Trans. on Components, Hybrids, and Manufacturing Technology*, vol. 16, 8, pp. 940–948, 1993.

[29] S.A. Dyer, *Survey of Instrumentation and Measurement*, Wiley, New York, 2001.

[30] W.C. Dunn, *Fundamentals of Industrial Instrumentation and Process Control*, Artech House Publishers, Boston, 2005.

Chapter 5

Optoelectronic and Photonic Sensors

5.1   Optoelectronic Microphone

Nanotechnology has revolutionized a number of classical applications by allowing the integration of optical sensing techniques into such applications. Advanced new products in this category include optical microphones, fingerprint readers, and highly sensitive seismic sensors. This section describes a highly advanced optical microphone that was introduced very recently. The microphone uses a charge-coupled device (CCD) as its sensing mechanism. This new type of microphone will be available commercially within a time span of 1 to 3 years.

5.1.1   Introduction and Principle of Operation

To overcome the well-known limitations of microphones that are based on electromagnetic transduction principles [1], a few attempts to use principles of optics to improve the basic microphone have appeared in the literature during the past decade [1–4]. Essentially, the optical technique that was successfully demonstrated is the use of laser interferometry to detect the minute vibrations of the microphone's diaphragm. Laser interferometry is capable of delivering excellent characteristics in very small packages [4]; however, the technique is quite expensive. Furthermore, the dynamic range of microphones that are based on laser interferometry is still not competitive with traditional microphones. In a simpler type of optical microphone [3], the interferometer is eliminated and only the intensity of the light that is reflected from the microphone's diaphragm is used for sensing the position. That type of design, however, suffers from very poor performance at frequencies above 10 kHz.

Very recently, a novel new optical microphone that is characterized by structural simplicity, low cost, very large dynamic range, and excellent frequency response characteristics was introduced. The fundamental principle behind the new microphone is illustrated in Figure 5.1. As shown in Figure 5.1(a), a conventional, thin metallic diaphragm vibrates in response to sound pressure. The diaphragm has a reflective pattern that is etched on its surface (the pattern used in the present prototype is a group of reflective circles). A group of light-emitting diodes (LEDs) are arranged underneath the diaphragm, as shown, to illuminate the reflective pattern. The image of the pattern is then captured by a CCD and is passed to a microprocessor for pattern recognition. Hence, pattern recognition is the distinctive new feature in this type of microphone. Figure 5.1(b) shows the typical picture of the pattern when the diaphragm is flexed away from the CCD. Figure 5.1(c) shows the perceived pattern when the diaphragm is flexed toward the CCD. With the availability of cheap, high speed electronics, the CCD can be easily sampled at a rate of 32 to 40 kHz (or the Nyquist rate for high-bandwidth music), and pattern recognition can be performed in real time to detect the instantaneous position of the vibrating diaphragm.

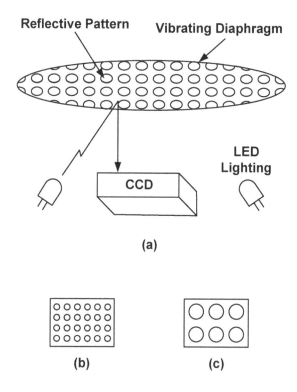

Figure 5.1: Principle of operation of the advanced optical microphone.

Figure 5.2: Main components of the microphone: the patterned metallic diaphragm, the circular LED array, a wide-angle fixed-focus lens, and the CCD (mounted on its own interface board). One US penny is shown in the photo for size comparison.

Figure 5.2 shows a photograph of the four main components of the prototype that was assembled by the author: the patterned metallic diaphragm, the circular LED array that is used for illuminating the diaphragm, the CCD, and a wide-angle fixed-focus lens that is used for focusing the image of the diaphragm on the CCD.

Figure 5.3 is a mechanical diagram that shows how the components were assembled to form a fully integrated microphone. It is to be pointed out that the wide-angle, fixed-focus lens that is used in the prototype can keep the diaphragm's image focused on the CCD even if the diaphragm moves by as much as a few millimeters (in actual microphony applications, movements of the diaphragm are typically on the order of nanometers or angstroms).

The CCD used in the present prototype is a low resolution (640×480 pixels) CCD that comes with the CMU CAM image processing board [4]. The CMU CAM is an off-the-shelf miniature image processing board that was chosen for the present prototype. A number of custom-modified versions of the board are available commercially. It is to be pointed out that an image subsample of only 32×32 pixels is actually used in the present application due to timing constraints on the small, low-cost hardware that was chosen (see Section 5.1.3 below). As shown in Figure 5.2, the CCD is mounted on an interface board for row/column scanning, and the interface board connects to the microprocessor board with a flexible connector. The LEDs that are used for illumination are miniature red LEDs, as CCD devices are known to be most sensitive to red light [5]. However, LEDs of any color can in principle be used in this application. The diaphragm that was used is a very thin metallic film with a

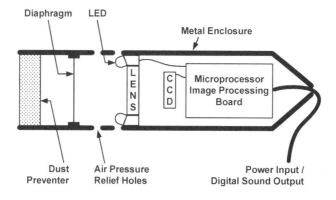

Figure 5.3: Mechanical diagram of the integrated microphone assembly.

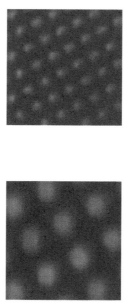

Figure 5.4: Pictures of the diaphragm as captured by the CCD: with the diaphragm flexed 0.5 mm away from the CCD (upper), and with the diaphragm flexed 0.5 mm toward the CCD (lower). The photographs are 32 × 32 pixels each.

thickness of approximately 0.01 mm and has a group of reflective circles/dots on one of its surfaces as shown (these types of decorated metallic films are widely available from numerous industrial suppliers).

Figure 5.4 shows two photographs that were captured by the CCD as the diaphragm was flexed 0.5 mm away from the CCD (upper) and 0.5 mm toward the

Table 5.1 **Comparison of the new microphone to the other known types of optical microphones [1–4].**

	Cost of Technology	Dynamic Range	Demonstrated Frequency Response Range	Total Harmonic Distortion
New microphone	Very low	20–140 dB SPL	20 Hz–16,000 Hz	Low ($< 1\%$)
Interferometry-based microphones	High	30–120 dB SPL	20 Hz–20,000 Hz	Low ($< 1\%$)
Light intensity-based microphones	Low	30–120 dB SPL	20 Hz–8,000 Hz	Low ($< 1\%$)

CCD (lower). The photographs are 32×32 pixels each (the actual image size used in the present application) and are substantially magnified in Figure 5.4. As pointed out above, movements of the diaphragm are typically on the order of angstroms or nanometers in actual microphony applications (substantially less in magnitude than the movements shown in Figure 5.4); however, the minor differences in the images, while they will not be visible to the naked eye, are quite distinguishable with the image processing algorithms. The image processing algorithm that is used in the present prototype is fully described in Sections 5.1.2 and 5.1.3 below.

A note is now in order concerning the limitations of microphones that are based on electromagnetic transduction principles. Many of these microphones suffer from nonuniform frequency response characteristics and harmonic distortion that can reach unacceptable levels. Some types of these microphones also have poor dynamic range. Optical microphones, such as the new device introduced in this section, mainly solve the first two of these problems (that is, they offer flat frequency response and extremely low harmonic distortion). The dynamic range, however, while not competitive with the best electromagnetic microphones, is quite acceptable for the majority of applications (the dynamic range is 20 to 140 db SPL in the current device). Table 5.1 gives a comparison of the main characteristics of the new microphone to those of other optical microphone technologies.

5.1.2 *Theory*

The instantaneous displacement of the diaphragm:

The vibrating metallic diaphragm in the present microphone can be modeled as a forced harmonic oscillator. The standard solution of the problem of the forced harmonic oscillator gives the instantaneous displacement $x(t)$ of the oscillator as [6,7]

$$x(t) = \frac{F_{max} \sin(\omega t + \phi)}{m\sqrt{(\omega_0^2 - \omega^2)^2 + \beta^2 \omega^2}} \tag{5.1}$$

where m is the mass of the oscillator (the diaphragm), $\omega_0 = \sqrt{k/m}$ is the natural frequency of oscillation, k is the spring constant, β is the damping constant, $F_{max} \sin \omega t$ is the sinusoidal force acting on the oscillator, and ϕ is the possible phase shift. It is to be pointed out that more sophisticated mathematical models for microphone diaphragms were developed in the past [13,14]. These models usually attempt to for-mulate/obtain the instantaneous displacement more accurately by coupling Eq. (5.1) with the numerical solution of the Navier–Stokes equations (to model the flow of air around structures that may be close to the diaphragm). Such elaborate mathematical models are needed for microphones such as the condenser microphone, where the diaphragm is very close to other components. However, for an isolated diaphragm, such as in the present application, Eq. (5.1) is usually quite sufficient for predicting the displacement of the diaphragm with a good degree of accuracy. The force that is acting on the diaphragm is equal to the acoustic pressure P multiplied by the surface area A of the diaphragm, hence

$$x(t) = \frac{A}{m\sqrt{(\omega_0^2 - \omega^2)^2 + \beta^2\omega^2}} P_{max} \sin(\omega t + \phi) \tag{5.2}$$

In the present application, the ratio A/m for the diaphragm is approximately 241 m^2/kg. The only two unknown constants in Eq. (5.2) are the natural frequency ω_0 and the damping constant β. ω_0 was determined experimentally by directly measuring it with a scanning laser vibrometer (see the experimental results in Section 5.1.4). By knowledge of ω_0, β can be determined from Eq. (5.2) if the displacement $x(t)$ is simply measured at the natural frequency; that is, if $\omega = \omega_0$,

$$\beta = \frac{A/m}{x_{max}\omega_0} P_{max} \tag{5.3}$$

and hence β can be determined by measuring the maximum displacement x_{max} for a certain applied acoustic pressure P_{max}. From the measurements, the natural (i.e., res-onant) frequency f_0 ($f_0 = \omega_0/2\pi$) was found to be equal to 4657 Hz for the present prototype. β was determined to be equal to 0.017 N.s/m, per unit mass. Details about the measurements are described in Section 5.1.4.

Detection of the instantaneous displacement via pattern recognition:
The instantaneous displacement $x(t)$ is determined by means of pattern recogni-tion in the present application. The algorithm subsequently uses Eq. (5.2) to "equal-ize" the measurement and obtain a flat frequency response (see Section 5.1.3). Nu-merous pattern recognition algorithms and approaches can be used in the present ap-plication, including those that are based on the powerful Fourier transform and neural network techniques. These sophisticated techniques, however, are also computation-ally intensive and hence cannot be practically used in a small handheld device. The approach that was chosen for the present prototype is a simple yet robust approach: the properties of the covariance matrix [16,18]. In 2-dimensional pattern recognition applications, the covariance matrix C is defined as

$$C = \frac{1}{M} \sum_{i=1}^{N} \sum_{j=1}^{N} (\vec{x}_i - \bar{x}_i)(\vec{x}_j - \bar{x}_j)^T \tag{5.4}$$

where M and N are the number of rows and columns, respectively, in the image matrix, $\vec{x}_i$ is a column vector, $\bar{x}_i$ is the mean value of $\vec{x}_i$, and T denotes a transpose. Usually, image luminosity data are normally distributed [16,18], with the probability density function (PDF)

$$P(\vec{x}_i) = \frac{1}{\sqrt{2\pi|C|}} \exp\left(-\frac{1}{2}(\vec{x}_i - \bar{x}_i)C^{-1}(\vec{x}_i - \bar{x}_i)^T\right) \tag{5.5}$$

If a certain vector $\vec{x}_i$ in the image is compared to the same vector in an image taken when the diaphragm is stationary, the ratio of the two corresponding PDFs will be

$$\frac{P(\vec{x}_i)}{P(\vec{x}_0)} = \exp\left(\frac{1}{2}(\vec{x}_0 - \bar{x}_0)C^{-1}(\vec{x}_0 - \bar{x}_0)^T\right.$$
$$\left. -\frac{1}{2}(\vec{x}_i - \bar{x}_i)C^{-1}(\vec{x}_i - \bar{x}_i)^T\right) \tag{5.6}$$

where $\vec{x}_0$ is the same vector in an image taken when the diaphragm is stationary. A "recognition function," $R(\vec{x}_i)$, can now be defined by taking the natural logarithm of the previous expression; that is,

$$R(\vec{x}_i) = \ln\frac{P(\vec{x}_i)}{P(\vec{x}_0)} =$$
$$\frac{1}{2}\left((\vec{x}_0 - \bar{x}_0)C^{-1}(\vec{x}_0 - \bar{x}_0)^T - (\vec{x}_i - \bar{x}_i)C^{-1}(\vec{x}_i - \bar{x}_i)^T\right) \tag{5.7}$$

Accordingly, since a positive displacement will result in a given vector $\vec{x}_i$ in the image having more luminosity than $\vec{x}_0$ (and, similarly, a negative displacement will result in $\vec{x}_i$ having less luminosity than $\vec{x}_0$), it can be immediately seen that the recognition function $R(\vec{x}_i)$ defined above will be positive for positive displacements of the diaphragm and negative for negative displacements. Figure 5.5 shows a typical plot of $R(\vec{x}_i)$ in the present application.

As can be seen from Figure 5.5, a displacement of less than 0.01 nm can be detected with the recognition function given by Eq. (5.7). This is not surprising, since far less sophisticated optical microphones (which are currently available commercially) offer the same performance [3]. Figures 5.6 and 5.7 show graphical representations of the covariance matrices that were calculated from the two images in Figure 5.4. The covariance matrices consist of real numbers, and the figures show blocks that correspond to those numbers (darker colors represent larger numbers in the matrix and lighter colors represent smaller numbers. Red represents the highest numbers). Of course, for displacements on the order of nanometers or less, the differences in the covariance matrices will be substantially smaller, but will still be distinguishable mathematically. In the approach described above for differentiating between images by using the recognition function, the only calculation that is computationally intensive is the calculation of the inverse of the covariance matrix (C^{-1}). This computation, however, is well within the capability of a small microprocessor for two reasons: first, the covariance matrix is a symmetric matrix, and second, the size of the image that is processed in the present application is quite small (32×32 pixels).

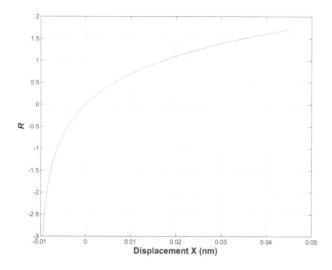

Figure 5.5: Recognition function *R* vs. diaphragm displacement in nm (calculated at 20 dB SPL). At 20 dB SPL, the recognition function can detect a displacement of 0.01 nm or less.

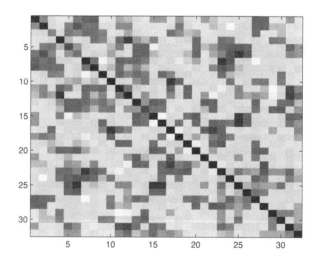

Figure 5.6: Graphical representation of the covariance matrix calculated from the upper image in Figure 5.4.

5.1.3 *Description of the Image Acquisition/Pattern Recognition Hardware and Software*

The image processing board that was chosen for the present application is the CMU CAM board [4]. This is a miniature image processing/pattern recognition board that

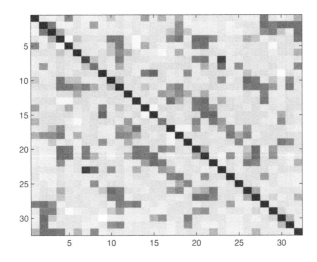

Figure 5.7: Graphical representation of the covariance matrix calculated from the lower image in Figure 5.4.

was originally designed for robotics applications. The latest versions of the board use a 32-bit microprocessor that runs at a clock frequency of 80 MHz. In this application, it is desirable to sample the CCD with a frequency in the range of 32 to 40 kHz (or the Nyquist rate for high-bandwidth music). By knowledge of the sampling rate, the wordlength, and the duration of a Read cycle of the microprocessor, the maximum number of bits that can be fetched by the microprocessor in one sample can be easily calculated. This number was determined to be approximately 10,000 bits at 32 kHz. The CCD that is used in conjunction with the microprocessor in this application has a total of 307,200 pixels, and, in grayscale mode, each pixel is represented by 8 bits. The total image is therefore represented by 2.4 million bits. Unfortunately, this far exceeds the maximum number of bits that can be processed by the hardware in one sample. In addition, while the microprocessor is based on parallel-processing architecture, it was found that some overhead time is required for image processing and decision-making. The practical size of the image that met such constraints was determined to be 32×32 pixels (or 8192 bits in grayscale mode). Hence, only a small subsample of the image provided by the CCD is used in this application. The sampling rate was also limited to 32 kHz. However, with more expensive and capable hardware, the sampling rate can be easily extended to 40 kHz.

Figure 5.8 shows a plot of the maximum displacement x of the diaphragm, as given by Eq. (5.2), for a maximum sound pressure (P_{max}) of 20 db SPL. Figure 5.9 shows the same plot for a maximum sound pressure of 60 db SPL (or normal speech). Clearly, the expected displacement at 20 SPL is a fraction of an angstrom (this is a common displacement for microphone diaphragms at low sound pressure levels). However, with 2^{8192} possible luminosity values, such minute displacements can be detected by the software. It was determined experimentally that an image size of 32

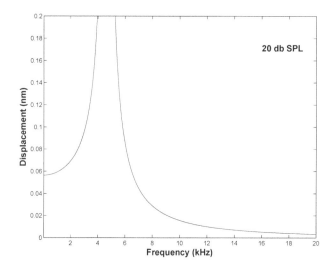

Figure 5.8: Maximum displacement of the diaphragm vs. frequency at a sound pressure of 20 db SPL.

× 32 pixels will meet a sensitivity of 20 db SPL, but a lower image size will not meet such sensitivity (for this reason, the dynamic range of the prototype described here starts at 20 db SPL). In the current implementation, each image sample of 32 × 32 pixels is treated as one vector of 1024 different numbers, and the recognition function R is computed for each sample by knowledge of the "base" image at zero displacement (see Eq. (5.7)). As Figure 5.5 shows, R is a simple logarithmic function of the instantaneous displacement $x(t)$ of the diaphragm (with the origin shifted to 0). Hence, once R is calculated, $x(t)$ is immediately calculated as $\exp(R)$. This step is finally followed by origin shifting and scaling.

To summarize, the algorithm compares two images that are separated by a very small displacement: a reference image where the displacement is equal to zero, and a second image for which the displacement needs to be estimated. The very minor differences in the images are detected by comparing the covariance matrices. A note is now in order concerning the calibration of the microphone (that is, how the exact displacement shown in Figure 5.5 was calculated). Calibration was performed inside an anechoic chamber by placing the microphone at a distance of a few inches from a loudspeaker to which a sinusoidal waveform was supplied. The sound pressure at the location of the microphone was measured with a Brüel and Kjaer Model 2236 sound pressure meter. By knowledge of P_{max} and the frequency used, $x(t)$ can be precisely calculated from Eq. (5.2) (see Figures 5.8 and 5.9). With the values of R being determined according to the procedure described above (with a frequency of 32,000 samples/sec), exact calibration of this sensing mechanism can be performed. Figure 5.5 shows the result of that calibration (performed at 20 db SPL).

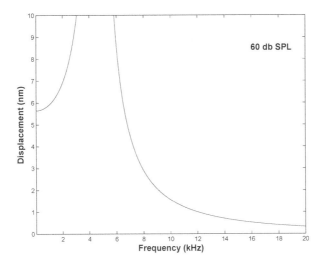

Figure 5.9: Maximum displacement of the diaphragm vs. frequency at a sound pressure of 60 db SPL.

As the reader might have expected, it was confirmed experimentally that image samples that are larger than 32×32 pixels will satisfy higher sensitivities (down to 0 db SPL), while smaller image sizes will not meet the sensitivity of 20 db SPL reported above. As indicated earlier, 32×32 pixels was determined to be the optimal image size that will simultaneously satisfy the two requirements of high performance and low cost in this new microphone. Figure 5.10 is a flowchart that shows how the code is structured and executed in the present application. As the flowchart shows, the last step in the code is the equalization step. Equalization is performed to obtain a flat frequency response and is a common procedure in digital microphones [1, 8, 9, 19]. Essentially, the computed instantaneous displacement $x(t)$ must be multiplied by the square root in the denominator of Eq. (5.2), $\sqrt{(\omega_0^2 - \omega^2)^2 + \beta^2 \omega^2}$, in order to eliminate the dependence on the frequency and make $x(t)$ a function of the acoustic pressure only. The only unknown in this quantity is the instantaneous angular frequency ω. To determine this quantity, a field programmable gate array (FPGA) was designed for estimating the frequency f in real time. The FPGA performs this function by fetching 256 successive samples of the calculated signal $x(t)$ and performing a fast Fourier transform (FFT) to determine its instantaneous frequency $f(t)$ (where the highest peak obtained in the power spectrum plot corresponds to the instantaneous frequency). The instantaneous frequency is then fed to the microprocessor for calculating the quantity under the square root. It is to be pointed out that it was necessary to develop a FPGA for this application (and hence perform the 256-point FFT in hardware rather than in software), since the FFT is known to be time consuming, and, accordingly, it was determined that the timing constraints would not be met in the present prototype if equalization had to be performed entirely in software.

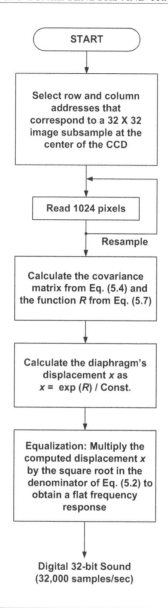

Figure 5.10: Flowchart of the code used in conjunction with the image processing board.

5.1.4 Experimental Results

Determination of the resonant frequency:

The resonant (natural) frequency of the diaphragm was determined by measuring the unit-step function response of the diaphragm with a scanning laser vibrometer

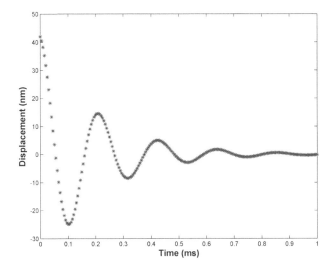

Figure 5.11: Unit-step function response of the diaphragm.

(Polytec model PSV-400). The step function was applied to the microphone acoustically, by using a loudspeaker that was placed directly in front of the microphone. The resonant frequency was found to be approximately 4657 Hz. The displacement data of the diaphragm, as measured by the vibrometer, is shown in Figure 5.11.

Probability of error as a function of the image size:

The probability of calculating the displacement $x(t)$ incorrectly was measured as a function of the image size. Figure 5.12 shows the results at 20 db SPL. Figure 5.13 shows the results at 60 db SPL. All the measurements were performed inside an anechoic chamber. As described earlier, the microphone was subjected to a sinusoidal sound pressure wave and a Brüel and Kjaer Model 2236 sound pressure meter was used as a reference for determining the acoustic pressure. Clearly, the probability of error increases substatinally as the size of the image matrix becomes smaller. Once again, an image size of 32 × 32 pixels was determined to be the optimal size that satisfies the two requirements of high performance and low hardware cost.

Frequency response characteristics and microphone's self-noise:

The frequency response of the microphone was also measured inside the anechoic chamber. Figure 5.14 shows a sample plot of a voice/music segment in the time domain, and Figure 5.15 shows the measured frequency response for such inputs. It is to be pointed out that the maximum frequency that was tested is 16 kHz (as indicated above, 16 kHz is the maximum frequency that is allowed in the current design). As can be concluded, the equalization procedure described earlier does indeed result in a flat frequency response.

The self-noise of the microphone was also measured and is shown in Figure 5.16. Figure 5.16 was obtained by summing the frequency components of the noise in each

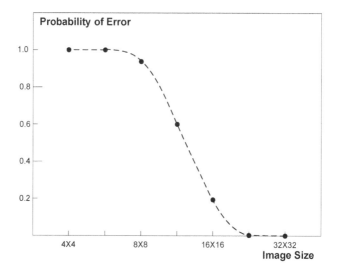

Figure 5.12: Probability of error as a function of the image size at 20 db SPL. (Horizontal axis shown with arbitrary scale.)

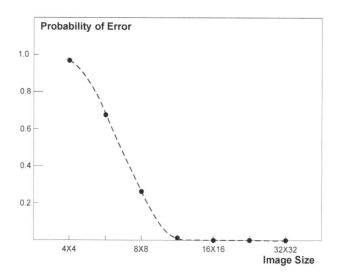

Figure 5.13: Probability of error as a function of the image size at 60 db SPL. (Horizontal axis shown with arbitrary scale.)

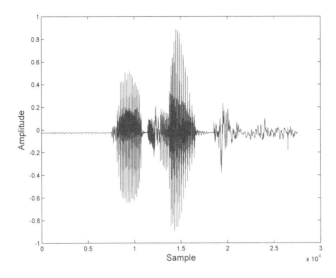

Figure 5.14: Time domain segment that shows one of the various voice/music inputs that were used for testing the frequency response of the microphone.

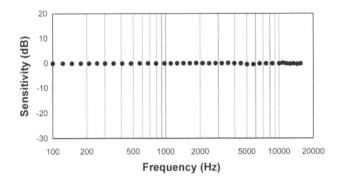

Figure 5.15: Frequency response of the microphone in the range of 100 Hz to 16 kHz.

bin and multiplying by an A-weighted filter in MATLAB®. As is well-known in the literature, the self-noise of most microphones is mainly a thermally induced noise (or thermal–mechanical noise). This noise is typically much higher than electrically induced noise (which may be due to circuitry or optical effects). As Figure 5.16 shows, the self-noise of the microphone (mostly thermal–mechanical) is slightly less than 20 dBA. It is to be pointed out that the noise floor of the anechoic chamber used is approximately −26 dBA.

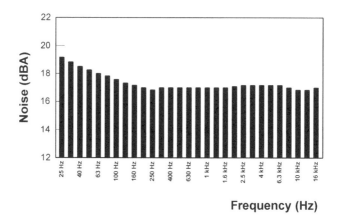

Frequency (Hz)

Figure 5.16: Microphone's noise spectrum in the range of 25 Hz to 16 kHz, in one-third octave bands. The noise floor of the anechoic chamber is approximately -26 dBA.

Measurement of the total harmonic distortion:

The total harmonic distortion (THD) was tested near the peak of the dynamic range of the microphone. At an acoustic pressure near 140 dB (which was measured with a Brüel and Kjaer Model 2239A sound pressure meter), the microphone was exposed to pure tones of frequencies in the range of 20 Hz to 16 kHz inside the anechoic chamber. The tests were repeated by using segments of speech and music. In each case, the digital signal generated by the image processing board was captured and analyzed in MATLAB with a FFT routine. When the power spectral density (PSD) was compared to the PSD of the original signal, the THD was found to be less than 0.5%. It was observed, however, that the THD grows significantly when the pressure exceeds 140 dB.

5.1.5 Conclusion

The new optical microphone introduced in this section features the same excellent qualities of optical acoustic sensors that are based on interferometry, with a much simpler design and at a fraction of the cost. Based on pattern recognition principles and highly integrated, high-speed electronics, the microphone captures image samples at a rate of 32 kHz and estimates the instantaneous displacement of the diaphragm from each sample. As indicated in this chapter, the use of off-the-shelf components limited the sampling rate and also the size of the image that can be processed in real time. Nevertheless, the prototype demonstrated excellent characteristics, including flat frequency response, very low harmonic distortion, and a dynamic range of 20–140 dB SPL. The intended applications of this new microphone are (a) any application where very large dynamic range is desired (for example, reporting from

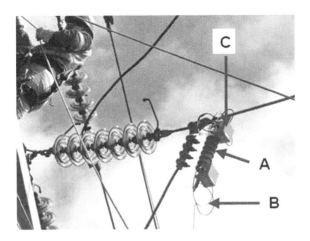

Figure 5.17: Hybrid optoelectronic sensor for current and temperature monitoring in overhead transmission lines, from [25].

a war zone), and (b) any application where very low harmonic distortion is desired (e.g., sound recording for analysis and forensics).

5.2   Other Optoelectronic and Photonic Micro Sensors

Other types of very advanced optoelectronic sensors were also introduced recently. Figure 5.17 shows an integrated hybrid optoelectronic sensor for current and temperature monitoring in overhead transmission lines. The sensor relies on nanostructured fiber optic lines for proximity sensing of parameters such as current and temperature (see [25]).

5.3   Quiz

1) New, advanced optical microphones are based on which scientific principle or theory?
Ans: The use of a charge-coupled device (CDD) and high-speed electronics to detect a minute mechanical displacement in real time.

2) True or false: The new CCD based optical microphone is characterized by structural simplicity, low cost, very large dynamic range, and excellent frequency response characteristics.
Ans: True.

References

[1] Ezzat G. Bakhoum and Marvin H.M. Cheng, Novel electret microphone, *IEEE Sensors Journal*, vol. 11, 4, pp. 988–994, 2011.

[2] Chen-Chia, Wang; Sudhir, Trivedi; Feng, Jin; V., Swaminathan; Ponciano, Rodriguez; and Narasimha, Prasad, High sensitivity pulsed laser vibrometer and its application as a laser microphone, *Applied Physics Letters*, vol. 94, art. no. 051112, 2009.

[3] Mauro Henrique, de Paula; Agleison Ramos, Omido; Daniel Freitas de Queiroz, Schumaher; and Hevelyne Henn, da Gama, Optical microphone: new results, *Review of Scientific Instruments*, vol. 75, 9, pp. 2863–2864, 2004.

[4] Michael L. Kuntzman, Caesar T. Garcia, A. G. Onaran, Brad Avenson, Karen D. Kirk, and Neal A. Hall, Performance and modeling of a fully packaged micromachined optical microphone, *IEEE J. of Microelectromechanical Systems*, vol. 20, 4, pp. 828–833, 2011.

[5] Alexander Paritsky and Alexander Kots, *Small Optical Microphone/sensor*, US Patent No. 6, 462, 808 B2, Oct. 8, 2002.

[6] Office of Undergraduate Research, Carnegie Mellon University, CMUcam: Open source programmable embedded color vision sensors, www.cmucam.org, 2012.

[7] Shigeyuki Ochi, *Charge Coupled Device Technlogy*, Gordon and Breach Pub., Amsterdam, The Netherlands, 1996.

[8] R. Feynman et al., *The Feynman Lectures on Physics*, Vol.1, Addison Wesley, Reading, MA, 1964.

[9] P.A. Tipler, *Physics*, Worth Pub., New York, 1986.

[10] Fukunaga, Keinosuke, *Statistical Pattern Recognition*, Academic Press, London, UK, 1990.

[11] J. Eargle, *The Microphone Book*, Elsevier/Focal Press, Burlington, MA, 2005.

[12] M. Gabrea, E. Grivel, and M. Najun, A single microphone Kalman filter-based noise canceller, *IEEE Signal Processing Letters*, vol. 6, 3, pp. 55–57, 1999.

[13] K. Watanabe, Y. Kurihara, T. Nakamura, and H. Tanaka, Design of a low-frequency microphone for mobile phones and Its application to ubiquitous medical and healthcare monitoring, *IEEE Sensors Journal*, vol. 10, 5, pp. 934–941, 2010.

[14] J.F. Sear and R. Carpenter, Noise-cancelling microphone using a piezoelectric plastics transducing element, *IEEE Electronics Letters*, vol. 11, 22, pp. 532–533, 1975.

[15] K. Ono, M. Matsumoto, and H. Naono, Noise-canceling microphone for video cameras, *IEEE Trans. on Consumer Electronics*, vol. 36, 3, pp. 635–641, 1990.

[16] F.V.B. de Nazare and M.M. Werneck, Hybrid optoelectronic sensor for current and temperature monitoring in overhead transmission lines, *IEEE Sensors Journal*, vol. 12, 5, pp. 1193–1194, 2012.

Chapter 6

Biological, Chemical, and "Lab on a Chip" Sensors

6.1 "Lab on a Chip" Sensors

Chemical and biological sensors have advanced tremendously during the past two decades thanks to the advances in nanotechnology and nanofabrication. The state of the art of chemical and biological sensors is an integrated sensor known as a "Lab on a Chip." The market for Lab on a Chip products is expected to reach $14 billion in 2018 [1]. The Lab on a Chip concept is currently believed to be the ultimate solution for poorly equipped medical facilities and mobile healthcare platforms that lack the capabilities of a fully equipped laboratory. The Lab on Chip solutions that currently exist can easily perform diagnostic operations such as microorganism detection and characterization, flow cytometry applications, polymerase chain reaction (PCR), in addition to many others.

A Lab on a Chip is essentially composed of three sections: actuation, sensing, and electronic interface circuitry. A block diagram of such a chip is shown in Figure 6.1

A scanning electron microscope (SEM) micrograph of an actual Lab on a Chip is shown in Figure 6.2. In the actuation section, the chip generates electrical or mechanical forces that act on the biological sample (cells/fluid). In the sensing section, sensors that are embedded in the chip measure responses that may be electrical, optical, thermal, or magnetic, and route the detected signals to the electronic section for final processing. The electronic interface circuitry finally performs traditional signal processing functions such as amplification and noise reduction.

Figure 6.3 shows a typical interface of a Lab on a Chip to a computer's data acquisition system.

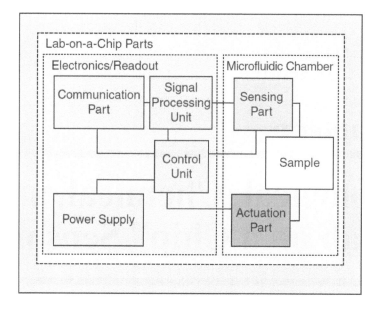

Figure 6.1: General structure of "Lab on a Chip," from Ghallab and Ismail [1].

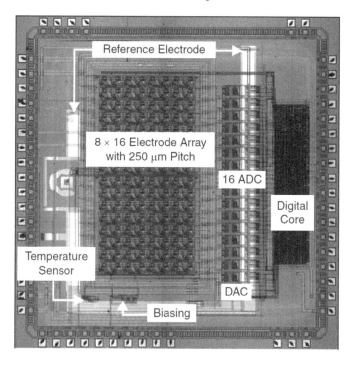

Figure 6.2: SEM micrograph of the "Lab on a Chip," from [2].

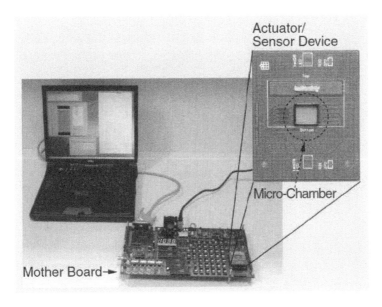

Figure 6.3: Typical interface of a "Lab on a Chip" to a computer data acquisition system, from [3].

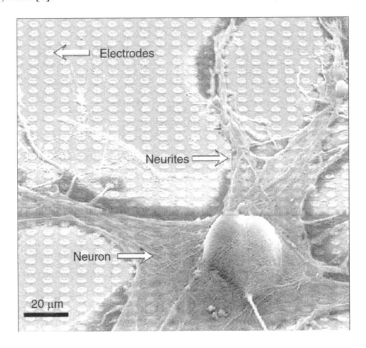

Figure 6.4: Neuron of a snail analyzed by a "Lab on a Chip," from [4].

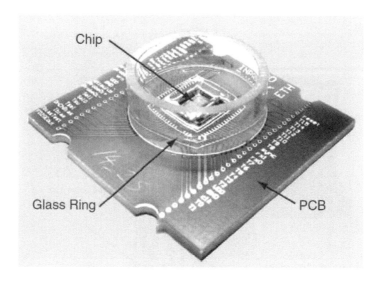

Figure 6.5: Typical packaging of a "Lab on a Chip," from [2].

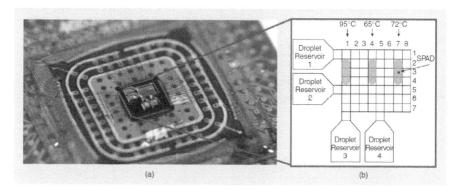

Figure 6.6: "Lab on a Chip" implemented directly on a printed circuit board, from [5].

Figure 6.4 shows a typical advanced application, where the neuron of a snail is tested for various responses by a Lab on a Chip.

Figures 6.5 and 6.6 show two different methods for packaging a Lab on a Chip.

6.2   Other Biochemical Micro- and Nano-Sensors

Other technologies for chemical and biological sensing have also been introduced recently. Figure 6.7 shows a CMOS on-chip sensor for measuring the dielectric constant of organic chemicals (see [6]).

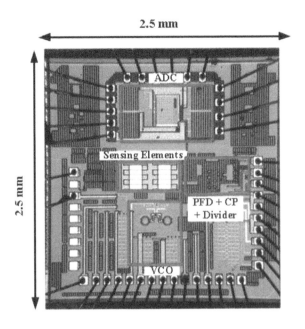

Figure 6.7: CMOS on-chip sensor for measuring the dielectric constant of organic chemicals, from [6].

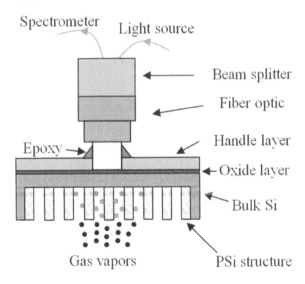

Figure 6.8: Porous silicon based sensor for chemical gas vapor detection, from [7].

Figure 6.8 shows a concept for a porous silicon based sensor for chemical gas vapor detection (see [7]).

6.3 Quiz

1) True or false: A Lab on a Chip sensor is, in many medical and analytical applications, an adequate replacement for a complete laboratory.
Ans: True.

2) True or false: A complete Lab on a Chip has recently been fabricated as an ordinary CMOS integrated circuit.
Ans: True.

References

[1] Y.H. Ghallab and Y. Ismail, CMOS based lab-on-a-chip: Applications, challenges and future trends, *IEEE Circuits and Systems Magazine*, 20 May 2014, pp. 27–47.

[2] F. Heer, S. Hafizovic, W. Franks, A. Blau, C. Ziegler, and A. Hierlemann, CMOS microelectrode array for bidirectional interaction with neuronal networks, *IEEE J. Solid State Circuits*, vol. 41, 7, 2006.

[3] G. Medoro, N. Manaresi, A. Leonardi, L. Altomare, M. Tartagni, and R. Guerrieri, A lab-on-a-chip for cell detection and manipulation, *IEEE Sens. J.*, vol. 3, 3, pp. 317–325, 2003.

[4] B. Eversmann et al., A 128 128 CMOS biosensor array for extracellular recording of neural activity, *IEEE J. Solid State Circuits*, vol. 38, 12, pp. 2306–2317, Dec. 2003.

[5] H. Norian, I. Kymissis, and K. L. Shepard, Integrated CMOS quantitative polymerase chain reaction lab-on-chip, in *Proc. Symp. VLSI Circuits* (VLSIC), June 12-14, pp. C220–C221, 2013.

[6] A.A. Helmy, Jeon, Hyung-Joon, Lo, Yung-Chung, A.J. Larsson, R. Kulkarni, Kim, Jusung, J. Silva-Martinez, and K. Entesari, A Self-sustained CMOS microwave chemical sensor using a frequency synthesizer, *IEEE J. Solid State Circuits*, vol. 47, 10, pp. 2467–2483, 2012.

[7] T. Karacali, U.C. Hasar, I.Y. Ozbek, E.A. Oral, and H. Efeoglu, Novel design of porous silicon based sensor for reliable and feasible chemical gas vapor detection, *IEEE J. of Lightware Technology*, vol. 31, 2, pp. 295–305, 2013.

Chapter 7

Electric, Magnetic, and RF/Microwave Sensors

Enormous advances in electric field, magnetic field, and RF/microwave sensors, driven by nanotechnology, have occurred recently. This chapter begins with an introduction to the most important of these applications: a very advanced magnetic field sensor.

7.1 Magnetic Field Sensors

A new type of solid-state magnetic field sensor that is similar in size to a Hall effect sensor and that offers a sensitivity that is approximately an order of magnitude better than a Hall effect sensor was recently introduced. The significant advantage of the new sensor, however, is that it consumes no power. The sensor consists of a radioactive β-particle source and a silicon p-n junction. If no magnetic field is applied, the β particles enter the p-n junction and generate a steady DC voltage. Under the influence of a magnetic field, however, the β particles (or secondary electrons generated therefrom) follow a curved path and miss the p-n junction, and the magnitude of the output voltage drops. The new sensor will be very advantageous in battery-powered consumer products applications since it consumes no power.

7.1.1 Introduction and Principle of Operation

Magnetic field sensing is an active research area [1–10, 13, 14, 16, 18, 19]. Recent advances include techniques for detection of the magnetic force that acts on MEMS components [2, 4], magnetic-tunnel junctions [6], ferro-fluids that are embedded in-

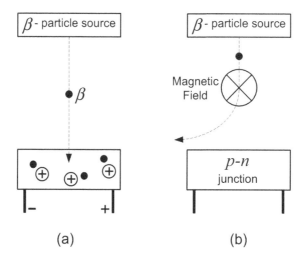

Figure 7.1: Fundamental principle of operation of the magnetic field sensor.

side optical fibers [7, 8]; giant magneto-resistance sensors [9, 19], in addition to improvements to the traditional Hall effect technique [13, 14]. At the present time, however, the magnetic field sensor of choice in most industrial applications is still the traditional Hall effect sensor [14, 15], due to its high sensitivity, small size, and low cost. The Hall effect sensor is also increasingly being used in portable electronic devices, such as digital cameras and smart tablets. Unfortunately, the traditional Hall effect sensor requires a current on the order of a few tens of mA to achieve its typical sensitivity, or about 10^{-4} T. This current requirement is undesirable in portable equipment, as it results in a shorter battery life. Although new "micro-power" Hall effect sensors have been introduced recently by a number of manufacturers [16], the magnetic field detection sensitivity is typically sacrificed as a result of the reduction in the current consumption of the device.

This chapter introduces a new type of solid-state magnetic field sensor that is characterized by:

■ Zero power consumption. The sensor, however, must be interfaced to signal conditioning circuitry like a traditional Hall effect sensor. Such circuitry can suitably be low-power circuitry.

■ Size that is comparable to a typical Hall effect sensor, with a magnetic field detection sensitivity that is at least an order of magnitude better than that of a typical Hall effect sensor (or about 10^{-5} T).

The new sensor depends on the beta-voltaic principle. Figure 7.1 shows the fundamental principle of operation of the new sensor. In Figure 7.1(a), a source of low-energy β particles is positioned above a p-n junction. As the β particles enter the p-n junction, electron-hole pairs are produced inside the junction, and a voltage appears across the terminals of the junction (the beta-voltaic principle is a well-known principle of physics and is described in numerous references [17, 24, 25]). In Figure 7.1(b), a magnetic field is applied in a direction that is perpendicular to the direction of motion of the β particles (or high-speed electrons). As a result, the β particles follow a curved path and miss the p-n junction, and the voltage across the terminals of the junction therefore drops in magnitude. Such a drop in voltage is related to the intensity of the applied magnetic field by a formula that is derived in Section 7.1.2.

Two important general remarks about the principle shown in Figure 7.1 are now in order:

■ The β-particle source used in the new sensor is tritium, with a half-life of 12.3 years. Such a half-life will be acceptable for portable consumer products, which typically have comparable lifetimes.[1]

■ β particles are high-speed electrons. Even if the source is a low-energy source such as tritium, the velocity of the β particles is substantially high (exceeds 10^7 m/s). According to the Lorentz force equation [18], however, such a high-speed particle cannot respond quickly to a magnetic field. For this reason, the fundamental principle shown in Figure 7.1 must actually be implemented with some important changes. Figure 7.2 shows the actual structure of the sensor.

As shown in Figure 7.2, the β-particle source is mounted immediately above a semiconductor body consisting of a p-type substrate and two n-doped regions. The two n-doped regions have the shape of concentric circles (as shown in the projection view in the lower part of the figure) and are physically separated from each other. Two separate p-n junctions therefore exist in the device. The p-region has one metal contact, as shown in the figure, while the two n-regions have separate metal contacts. β particles emitted from the decay of tritium have a maximum energy of 18.6 keV and an average energy of 5.7 keV [17, 24, 25]. These energetic β particles enter the p-type region, where they lose their kinetic energy within a distance of less than 1 μm (see Section 7.1.2). In the process, each β particle generates numerous electron-hole pairs (or EHPs) [17, 24, 25]. If no magnetic field is applied, the newly created EHPs drift toward the depletion region in the center, where the internal electric field in the depletion region forces the electron toward the n-type material and forces the hole back into the p-type material. As a result, a voltage V_1 appears across the terminals of the inner p-n junction, while the voltage V_2 across the outer p-n junction will be equal to 0. If, however, a magnetic field is applied in a direction that is perpendicular to the direction of motion of the β particles, the newly created EHPs (which have slow, nonrelativistic velocities) follow curved paths and accumulate on the sides of the p-substrate, as shown in the figure.[2] Accordingly, the outer p-n junction receives

---

[1] Tritium was chosen for the present application for technical (see Section 3) as well as safety reasons. Tritium, in small quantities, is approved for use in consumer products.

[2] In reality, the path of an electron inside a semiconductor is very irregular, due to the numerous elastic and inelastic collisions that the electron encounters within the lattice. However, the overall effect of the magnetic field on the path will be as shown in Figure 7.2.

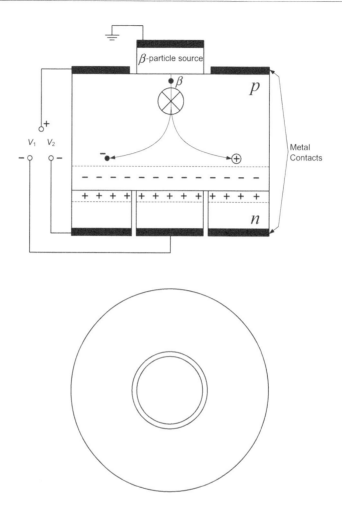

Figure 7.2: Cross-sectional view of the actual sensor (upper) and a projection view of the two n-doped regions (lower).

a larger portion of the EHPs and the voltage V_2 increases, while V_1 falls. The interface circuit of the sensor is designed to detect the voltage differential $V_2 - V_1$.

Figure 7.3 shows a photograph of the two main components of the sensor: the tritium β-particle source and the two concentric silicon p-n junctions. The component consisting of the two p-n junctions has a diameter of 12 mm and a total thickness of approximately 1 mm. Additional details about the procedure that was followed in the construction of the two components and the assembly of the prototype are given in the following sections.

Tritium β Source

p-n Junctions

Figure 7.3: The two components of the sensor: the tritium β source and the two concentric silicon p-n junctions. The diameter of the larger component is 12 mm. One US penny is shown in the photograph for size comparison.

7.1.2   Theory

Range of the β particles in silicon:

β particles emitted from the decay of tritium have a maximum energy of 18.6 keV and an average energy of 5.7 keV. The maximum range of penetration of a β particle into any material can be found from the Katz–Penfold formula [19, 20]

$$R_{max} = 0.412 E_{\beta}^{1.265-0.0954\ln(E_{\beta})}, \tag{7.1}$$

where R_{max} is the material-independent maximum range (in g/cm^2) and E_{β} is the initial energy of the β particle (in MeV). The maximum penetration distance in silicon will therefore be given by

$$R(\text{silicon}) = \frac{R_{max}}{\rho}, \tag{7.2}$$

where ρ = 2.33 g/cm^3 is the density of silicon [19, 20]. For an average kinetic energy E_{β} = 5.7 keV, the formula predicts a penetration depth in silicon of 0.2 μm. The penetration depth is therefore very small. However, each β particle generates numerous EHPs during its short journey, as is well-known from the theory of betavoltaics [17, 24, 25].

Bending radius of the generated free electrons:

The focus now is on the free electrons that will be generated in the p-region of the junction (see Figure 7.2). We shall obtain expressions for the minimum initial kinetic energy of the electron, its final kinetic energy, and the radius of the arc along its trajectory. These expressions will then be used to calculate the expected displacement of the electron in the horizontal direction. The electron mainly loses kinetic energy as it encounters inelastic collisions inside the material. Kinetic energy is also lost due to the Lorentz forces that will be acting on the electron in the direction opposite to the direction of motion. This last portion of the kinetic energy is the only portion that can be practically calculated, and we shall designate it as the "minimum" kinetic energy that will be needed in order for the electron to reach the depletion region. This minimum kinetic initial energy will be given by

$$\frac{1}{2}mv_0^2 = \int F \, dl \qquad (7.3)$$

where F is the Lorentz force, given by $F = q(\vec{E} + \vec{v} \times \vec{B})$. Here, as usual, m is the electron's mass, $\vec{v}$ is the electron's initial velocity, q is the electron's charge, $\vec{E}$ is the steady-state electric field that will be present between the terminals of the device (see Figure 7.2), $\vec{B}$ is the magnetic flux density of the applied magnetic field, and l designates the path of integration. The above equation can therefore be written as

$$\frac{1}{2}mv_0^2 = q \int \vec{E} \cdot dl + q \int (\vec{v} \times \vec{B}) \cdot dl \qquad (7.4)$$

Since the magnetic force is always perpendicular to the path, it can be immediately seen that the second integral vanishes, and the above equation reduces to

$$\frac{1}{2}mv_0^2 = qV, \qquad (7.5)$$

where V is the steady-state electrostatic potential, or voltage, between the terminals of the device. From Eq. (7.5), the minimum initial velocity v_0 will be given by

$$v_0 = \sqrt{\frac{2qV}{m}} \qquad (7.6)$$

The terminal (or final) kinetic energy of the electron as it approaches the depletion region can be found as follows: at steady-state (where electrons finally stop crossing the depletion region due to saturation of the p-n junction), the sum of the electron and hole currents must be equal to zero (see Figure 7.2). By using the well-known equation [21, 22]

$$J = nqv_e \qquad (7.7)$$

where J is the current density, n is the free-electron density in the p-region, and v_e is the terminal velocity of the electron, we must therefore conclude that

$$nqv_e + pqv_h = 0 \qquad (7.8)$$

where p is the hole density and v_h is the hole's terminal velocity. Hence,

$$v_e = -\frac{p}{n}v_h \qquad (7.9)$$

Holes will be directly driven by the steady-state electric field present between the terminals of the device, and v_h will be therefore given by [21, 22]

$$v_h = \mu_h E \tag{7.10}$$

where μ_h is the hole's mobility. Hence,

$$
\begin{aligned}
v_e &= -\frac{p}{n}\mu_h E \\
&= -\frac{p}{n}\mu_h\left(\frac{V}{L}\right) \tag{7.11}
\end{aligned}
$$

where L is the thickness of the device. In this application, electrons and holes are created in pairs. Since only these mobile carriers contribute to the current in the device, $n = p$ (in other words, the stationary holes present in the p-region do not contribute to the current and hence are not accounted for). Accordingly,

$$v_e = -\mu_h\left(\frac{V}{L}\right) \tag{7.12}$$

The negative sign merely indicates that the direction of motion of the electrons will be opposite to that of holes.

Typically, the loss of kinetic energy inside an atomic lattice is exponential [21, 22]; i.e., the kinetic energy of the electron takes the form

$$\frac{1}{2}mv^2 = \frac{1}{2}mv_0^2\, e^{-\alpha l} \tag{7.13}$$

where α is an attenuation constant. Since the total length l of the electron's path (before reaching the depletion region) is only slightly different from the overall thickness L of the device, then, to a very good approximation,

$$\alpha \approx -\frac{1}{L}\ln\left(\frac{v_e^2}{v_0^2}\right) \tag{7.14}$$

where v_0 and v_e are the final and the terminal velocities of the electron as previously indicated. Equation (7.13) therefore allows the calculation of the velocity v at any point along the electron's path by knowledge of v_0, v_e, and the arc length l. We are finally ready to obtain an expression for the bending radius of the electron's path. Along its path, the electron is in equilibrium due to the equality of the magnetic and the centrifugal forces, that is,

$$qvB = \frac{mv^2}{R} \tag{7.15}$$

where R is the arc's radius and where only the magnitudes are shown in the equation. Hence, from the above equation, and by using the results of Eqs. (7.13) and (7.14), R will be given by

$$
\begin{aligned}
R &= \frac{mv}{qB} \\
&= \frac{mv_0}{qB}\exp\left[\frac{l}{L}\ln\left(\frac{v_e}{v_0}\right)\right] \tag{7.16}
\end{aligned}
$$

The bending radius (or radius of curvature) can therefore be calculated at any point by knowledge of the initial velocity v_0, the final velocity v_e, and the arc length l. Since only the minimum value of v_0 can be estimated, then the radius of curvature so calculated will be the minimum radius of curvature. However, the final value of that radius can actually be determined with good accuracy: it can be immediately seen from the above equation that when $l = L$, the final radius of curvature will be given by

$$R = \frac{mv_e}{qB} \tag{7.17}$$

Deviation of the electron's path in the horizontal direction:
Starting with the expression for the radius of curvature given by Eq. (7.16), we can obtain another expression that describes the deviation of the electron's path in the horizontal direction. The path in this application is a typical logarithmic spiral[3] [23]. Any point on the logarithmic spiral is described by the following coordinates [23]:

$$\begin{aligned} x(\theta) &= a e^{b\theta} \cos\theta \\ y(\theta) &= a e^{b\theta} \sin\theta \end{aligned} \tag{7.18}$$

where a, b are constants and θ is a parameter along the curve. The arc length for any curve is given by the well-known equation [24]

$$l = \int_0^\theta \sqrt{[x'(\theta)]^2 + [y'(\theta)]^2} \tag{7.19}$$

By differentiating $x(\theta)$ and $y(\theta)$, substituting in Eq. (7.19), and carrying out the integration, we obtain

$$l = \frac{a\sqrt{b^2 + 1}}{b} \left[\exp(b\theta) - 1\right] \tag{7.20}$$

Now by writing θ as a function of l, we have

$$\theta = \frac{1}{b} \ln\left[1 + \frac{bl}{a\sqrt{b^2 + 1}}\right] \tag{7.21}$$

Since the exponent b must be a negative number, the maximum value of the arc length, l_{max}, according to Eq. (7.20), is

$$l_{max} = -\frac{a\sqrt{b^2 + 1}}{b} \tag{7.22}$$

By substituting from Eqs. (7.21) and (7.22) into Eq. (7.18), we obtain

$$\begin{aligned} x(l) &= a\left[1 - \frac{l}{l_{max}}\right] \cos\left(\frac{1}{b} \ln\left[1 - \frac{l}{l_{max}}\right]\right) \\ y(l) &= a\left[1 - \frac{l}{l_{max}}\right] \sin\left(\frac{1}{b} \ln\left[1 - \frac{l}{l_{max}}\right]\right) \end{aligned} \tag{7.23}$$

---

[3] As the electron travels inside the semiconductor lattice, it loses kinetic energy exponentially (see Eq. (7.13)). At the same time, it follows a curved path because of the applied magnetic field. The combination of the two effects results in a curve where the radius of curvature is exponentially shrinking or a logarithmic spiral curve.

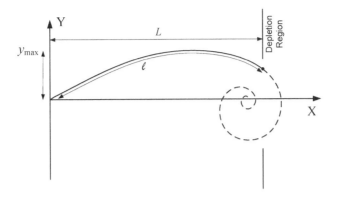

Figure 7.4: Simplified geometry of the logarithmic spiral curve in the present problem (the curve is rotated 90° for clarity).

An important property of the logarithmic spiral is that the arc length l at any point is equal to the radius of curvature R at that point [23]. Here, $l_{max} = R_0$, where R_0 is the radius of curvature at the beginning of the path. The above parameterized equations can therefore be written as

$$x(l) = a \left[1 - \frac{l}{R_0}\right] \cos\left(\frac{1}{b} \ln\left[1 - \frac{l}{R_0}\right]\right)$$

$$y(l) = a \left[1 - \frac{l}{R_0}\right] \sin\left(\frac{1}{b} \ln\left[1 - \frac{l}{R_0}\right]\right) \tag{7.24}$$

Since $l_{max} = R_0$, then, from Eqs. (7.16) and (7.22), we have

$$R_0 = \frac{mv_0}{qB} = -\frac{a\sqrt{b^2 + 1}}{b} \tag{7.25}$$

A second equation in the two unknowns a and b can now be found as follows: if the electron does reach the depletion region, then, near the end of the path,

$$x(l) \approx x(L) \approx L \tag{7.26}$$

as can be concluded from Figure 7.4 (it is to be pointed out that high accuracy in the estimation of the electron's coordinates is not necessary, as the analysis will prove to be in very close agreement with the experimental results. See Section 7.4.7). From Eqs. (7.24) and (7.26) we have

$$x(L) \approx L \approx a \left[1 - \frac{L}{R_0}\right] \cos\left(\frac{1}{b} \ln\left[1 - \frac{L}{R_0}\right]\right) \tag{7.27}$$

Since the thickness of the device L is substantially smaller than the initial radius of curvature R_0 (here, $R_0 = 33$ cm or higher, according to Eq. (7.25)), then the ratio

L/R_0 is negligible and the cosine term in the above equation is approximately equal to unity. Accordingly, to a very good approximation,

$$a \approx L \tag{7.28}$$

Equation (7.25) can now be rearranged to give

$$b = \pm \frac{a}{\sqrt{R_0^2 - a^2}} \tag{7.29}$$

where the sign in Eq. (7.29) must be a negative sign. From the last two equations we now immediately conclude that

$$b \approx -\frac{L}{R_0} \tag{7.30}$$

The objective now is to estimate the maximum deviation y_{max} (which, in this problem, is actually the horizontal deviation of the electron's path). Maximum deviation occurs at $dy/dl = 0$. By taking the derivative of the expression in Eq. (7.24) with respect to l and setting the result equal to 0, we obtain

$$l = R_0 \left[1 - \exp\left(b \tan^{-1} \frac{R_0}{L} \right) \right] \tag{7.31}$$

By knowledge of the value of l at the maximum deviation point (see Figure 7.4), the maximum deviation y_{max} can be found from Eq. (7.24). This is the final result that is needed in order to design such a device correctly. A simple calculation (see Section 7.4.7) shows that this expected deviation will be approximately equal to 0.5 mm for an applied magnetic flux density of 10^{-5} T. The numerical value of the deviation y_{max} is important because the voltage that will be produced by the inner p-n junction will be linearly proportional to the area of the junction that is exposed to the β-particle flux, which is in turn linearly proportional to the deviation y_{max}. Accordingly, the behavior of the voltage produced by the inner p-n junction should follow Eq. (7.24), with an unknown proportionality constant that is to be determined experimentally.

7.1.3 Manufacturing and Assembly of the Prototype Sensor

The tritium beta source (see Figure 7.3) was manufactured with relatively simple laboratory equipment. Tritium is the only β-particle source that is suitable for this application because the maximum energy of the emitted β-particles is 18.6 keV, which is just below the threshold of energy at which damage to the atomic lattice in silicon would occur [25–27] (that is, the possible displacement of an atom from the lattice due to the inelastic collision with an energetic β-particle). The small disk shown in Figure 7.3 is itself a disk of crystalline silicon of a thickness of 0.5 mm. Silicon, once again, was determined in recent research investigations to be an ideal material in which gaseous tritium can be embedded [28]. The process of embedding of tritium (or, "tritiation," as commonly known in the physics community) is a relatively simple process in which the crystalline silicon disk is exposed to tritium gas at a pressure of about 120 atm and a temperature in the range of 250–300°C [28]. The sample is typically exposed for several days. The tritium (in atomic form) diffuses inside the atomic silicon lattice up to a depth of approximately 10 nm [28, 29]. The concentration of the radioisotope in the sample is typically in the range of 5–10 mCi/cm². The tritium thus embedded in the silicon sample is usually found to be very stable at all temperatures below 300°C (above that temperature, the tritium may "evolve" and disappear from the silicon lattice).

The component consisting of two concentric p-n junctions, shown in Figure 7.3, was manufactured to order by a specialized semiconductor company in Taiwan. The prototype was finally assembled as shown in Figure 7.2.

7.1.4   Numerical Data and Experimental Results

Basic data:

The prototype sensor shown in Figure 7.3 was experimentally tested and was found to have the following parameters:

- Concentration of radioisotope in the source silicon layer: 8.64 mCi / cm²

- Nominal voltage across the inner p-n junction: 0.8 V

- Max current at rated voltage: 2.5 nA

- Minimum detectable magnetic flux density B: approximately 10^{-5} T

Figure 7.5 shows the voltage V_1 that appears across the inner p-n junction as a function of time (with no applied magnetic field). The voltage V_2 across the outer p-n junction was determined to be practically equal to 0 while no magnetic field is present. It is to be pointed out that the magnetic flux density of the earth is typically in the range of 25–65 μT, which is above the threshold of sensitivity of the sensor. Therefore, the magnetic field of the earth was shielded by placing the experimental setup inside a nickel-iron enclosure.

As Figure 7.5 shows, noise with a magnitude of about 3.5% is present in the voltage produced. This noise is due to the normal fluctuation in the β-particle flux that is reaching the p-n junction (at lower voltages, the noise was found to still be about 3.5% of the magnitude of the voltage produced). The bandwidth of the noise was found to be approximately 500 Hz.

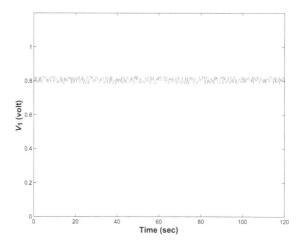

Figure 7.5: The voltage V_1 produced by the inner p-n junction as a function of time (with no applied magnetic field). The magnetic field of the earth was shielded by placing the experimental setup inside a nickel-iron enclosure.

Response of the sensor to DC magnetic fields:

A magnetic flux with a density B ranging from 10^{-6} T to 0.1 T was applied to the sensor. The magnetic field was generated with a simple, custom-made electromagnet that was placed in the vicinity of the sensor.[4] Once again, the magnetic field of the earth was shielded by placing the experimental setup inside a nickel-iron enclosure. The experimental setup is shown in Figure 7.7. Figure 7.6 shows the measured voltages V_1 and V_2 across the inner and outer p-n junctions, respectively, as the field was increased from 10^{-6} T to 0.1 T. The theoretically predicted value of V_1 was also calculated by calculating y_{max} from Eq. (7.24) and assuming that V_1 will be linearly proportional to that deviation. The theoretical and the experimental results are both shown in Figure 7.6. The data in Figure 7.6 was collected at room temperature (25°C).

From Figure 7.6, it is of course apparent that the voltage will be a nonlinear function of the applied magnetic flux density (generally, magnetic field sensors are non-linear devices). In practical applications, the magnetic flux density can be inferred from the measured voltage (V_1 or V_2) by inspection of the corresponding curve in Figure 7.6. Alternatively, in an automatic application (where the voltage must be connected to an interface/signal conditioning circuit), a "lookup table," which may be stored in a microprocessor, can be used for estimating the value of B. Figure 7.8 shows a typical plot of the difference $V_1 - V_2$ of the two voltages. It is to be observed,

---

[4]The magnetic flux density was varied by 5 orders of magnitude by simply varying the current in the electromagnet over such a range (since B in any electromagnet is directly proportional to the current [18]). More specifically, the current was varied from 2 μA to 200 mA in the experimental procedure described here.

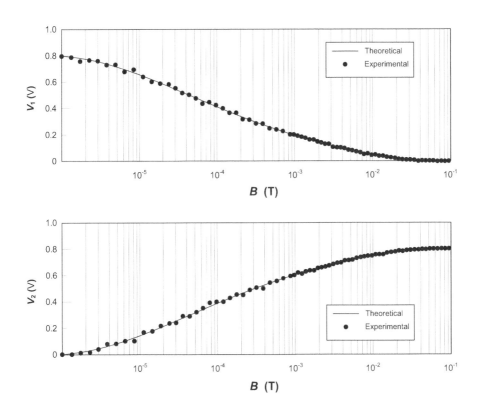

Figure 7.6: Voltages measured across the inner and outer p-n junctions as a function of magnetic flux density, at a temperature of 25°C.

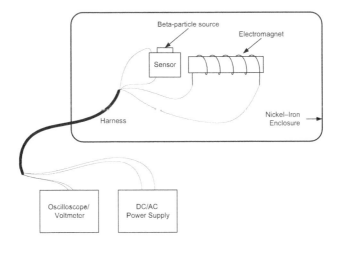

Figure 7.7: Diagram of the experimental setup for testing the sensor.

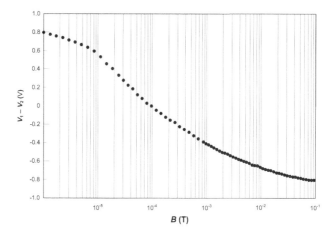

Figure 7.8: The voltage difference $V_1 - V_2$ as a function of magnetic flux density, at a temperature of 25°C.

however, that the sensor is inherently not a differential sensor (i.e., the "output" can be taken to be either V_1 or V_2 only).

Response to AC magnetic fields:

The tests described above were repeated with AC fields.[5] Figure 7.9 shows the response of the sensor to an alternating field with a peak value of 0.1 T and a frequency of 600 Hz. The low frequency was chosen for better visualization; however, the sensor was tested at frequencies of up to 100 MHz and the shape of the waveform was the same.[6] The response in Figure 7.9 can be understood as follows: as the magnetic flux density increases in value, the EHP current through the inner p-n junction is "chocked off" until it reaches 0 at the point of maximum applied field (0.1 T). The voltage across the inner junction increases again to 0.8 V as the field weakens, and, as the field reverses polarity, the EHP current decreases once again and the same cycle is repeated. The measurement shown in Figure 7.9 was taken at room temperature (25°C).

Temperature effect on the response of the sensor:

The sensor and the magnetic field source were placed inside a controlled-temperature chamber, and the tests described above were repeated. Figure 7.10 shows the results of two tests that were conducted with a DC field at temperatures of -40°C and $+120$°C. Clearly, lower temperatures have only a minimal effect on the sensitivity of the sensor, while a significant deviation in the output voltage occurs at high

---

[5]The AC signal was generated with an Agilent model N9310A signal generator. It was interfaced to the electromagnet with a simple transistor driver circuit. The output voltage was observed with an oscilloscope.

[6]There is a delay, however, in the response of the sensor at high frequencies, since electrons take time to travel through the silicon lattice.

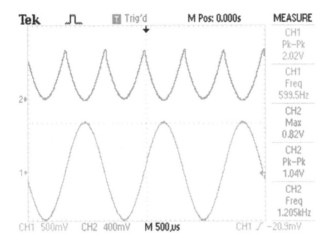

Figure 7.9: Oscilloscope plot of the voltage V_1 across the inner p-n junction (upper trace) and of the applied magnetic field (lower trace) at a frequency of 600 Hz.

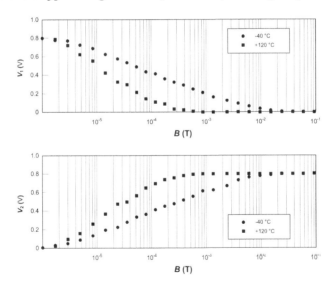

Figure 7.10: Voltages measured across the inner and outer p-n junctions as a function of magnetic flux density, at temperatures of $-40°$C and $+120°$C.

temperatures (above $+80°$C). The suitable operating range of the sensor is therefore $-40°$C to $+80°$C (this is typically the preferred operating range of a silicon p-n junction). Outside that range, an electronic circuit will be needed to compensate for the temperature dependence of the sensor.

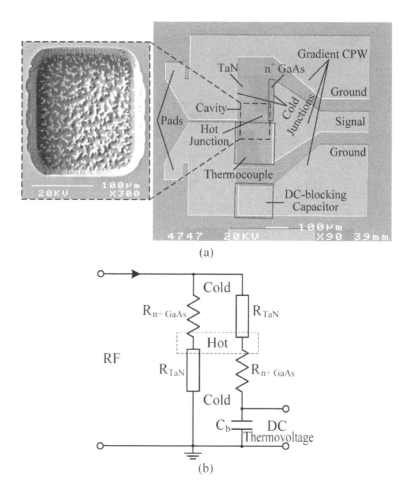

(a)

(b)

Figure 7.11: A thermocouple-based self-heating RF power sensor with GaAs MMIC-compatible micromachining technology, from [30].

7.1.5 Conclusion

The new magnetic field sensing technology introduced in this chapter will be very advantageous in battery-powered consumer products applications since it requires no power. The sensor described is comparable in size to a Hall effect sensor and offers a sensitivity that is at least an order of magnitude better than a typical Hall effect sensor. The radioactive β-particle source that is used in the sensor, namely, tritium, exists in numerous other consumer products (such as toys and glow-in-the-dark products) and is quite safe in small quantities.

One disadvantage of the new technology is the cost, since it is more complex than the Hall effect based sensing technology. With mass production, however, the cost can be substantially reduced. The sensor is expected to become available commercially within a time span of 1 to 3 years.

7.2 Other Important Electromagnetic/RF Micro- and Nano-Sensors

The field of RF, electric, and magnetic field sensing is advancing rapidly. Figure 7.11 shows an SEM microview of a self-heating radio frequency (RF) power sensor which senses RF power by detecting the heat generated in the sensor by means of two thermocouples.

7.3 Quiz

1) True or false: Miniature magnetic field sensors with ultrahigh sensitivity depend on the Hall effect.
Ans: False. These new advanced sensors use the beta-voltaic effect.

2) True or false: The advanced new magnetic field sensors that use the beta-voltaic effect include a radioactive beta-particle source.
Ans: True.

References

[1] G. Wu, D. Xu, B. Xiong, D. Feng, and Y. Wang, Resonant magnetic field sensor with capacitive driving and electromagnetic induction sensing, *IEEE Electron Device Letters*, vol. 34, 3, pp. 459–461, 2013.

[2] Y. Hui, T.X. Nan, N.X. Sun, and M. Rinaldi, MEMS resonant magnetic field sensor based on an AlN/FeGaB bilayer nano-plate resonator, *IEEE 26th International Conference on Micro Electro Mechanical Systems (MEMS)*, pp. 721–724, 2013.

[3] I. Bolshakova, I. Vasilevskii, L. Viererbl, I. Duran, N. Kovalyova, K. Kovarik, Y. Kost, O. Makido, J. Sentkerestiov, A. Shtabalyuk, and F. Shurygin, Prospects of using in-containing semiconductor materials in magnetic field sensors for thermonuclear reactor magnetic diagnostics, *IEEE Transactions on Magnetics*, vol. 49, 1, pp. 50–53, 2013.

[4] Sal M. Domnguez-Nicols, R. Jurez-Aguirre, P.J. Garca-Ramrez, Herrera-May, and L. Agustin, Signal conditioning system with a 4–20 mA output for a resonant magnetic field sensor based on MEMS technology, *IEEE Sensors Journal*, vol. 12, 5, pp. 935–942, 2012.

[5] J. Chen, M.C. Wurz, A. Belski, and H. Lutz, Designs and characterizations of soft magnetic flux guides in a 3-D magnetic field sensor, *IEEE Transactions on Magnetics*, vol. 48, 4, pp. 1481–1484, 2012.

[6] P. Wisniowski, J. Wrona, T. Stobiecki, S. Cardoso, and P.P. Freitas, Magnetic tunnel junctions based on out-of-plane anisotropy free and in-plane pinned layer structures for magnetic field sensors, *IEEE Transactions on Magnetics*, vol. 48, 11, pp. 3840–3842, 2012.

[7] A. Candiani, A. Argyros, R. Lwin, S. Leon-Saval, G. Zito, S. Selleri, and S. Pissadakis, A ferrofluid infiltrated polymeric microstructured optical fiber sensor for magnetic field measurements, *IEEE Photonics Conference (IPC)*, pp. 741–742, 2012.

[8] R. Picos, J. Font, E. Garcia, A. Pineda, and J. Cesari, Design and experimental characterization of a magnetic field sensor, *IEEE 8th International Caribbean Conference on Devices, Circuits and Systems (ICCDCS)*, pp. 1–4, 2012.

[9] L. Xianli, D. Hui, and H. Chunyang, A novel magnetic field sensor based on the combination use of microfiber knot resonator and magnetic fluid, *IEEE International Conference on Condition Monitoring and Diagnosis (CMD)*, pp. 111–113, 2012.

[10] B. Li, A. Morsy, and J. Kosel, Optimization of autonomous magnetic field sensor consisting of giant magnetoimpedance sensor and surface acoustic wave transducer, *IEEE Transactions on Magnetics*, vol. 48, 11, pp. 4324–4327, 2012.

[11] Linghao Cheng, Jianlei Han, Zhenzhen Guo, Long Jin, and Bai-Ou Guan, A novel miniature magnetic field sensor based on Faraday effect using a heterodyning fiber grating laser, *IEEE Photonics Global Conference (PGC)*, pp. 1–4, 2012.

[12] P. Childs, A. Candiani, and S. Pissadakis, Optical fiber cladding ring magnetic field sensor, *IEEE Photonics Technology Letters*, vol. 23, 13, pp. 929–931, 2011.

[13] A. Abderrahmane, S. Koide, S. Sato, T. Ohshima, A. Sandhu, and H. Okada, Robust Hall effect magnetic field sensors for operation at high temperatures and in harsh radiation environments, *IEEE Transactions on Magnetics*, vol. 48, 11, pp. 4421–4423, 2012.

[14] S.A. Dyer, *Survey of Instrumentation and Measurement*, Wiley, New York, 2001.

[15] W.C. Dunn, *Fundamentals of Industrial Instrumentation and Process Control*, Artech House Publishers, Boston, 2005.

[16] Maria-Alexandra Paun, Jean-Michel Sallese, and Maher Kayal, Hall effect sensors design, integration and behavior analysis, *Journal of Sensor and Actuator Networks*, vol. 2, pp. 85–97, 2013.

[17] P. Rappaport, The electron-voltaic effect in p-n junctions induced by beta-particle bombardment, *Physical Review*, vol. 93, 1, pp. 246–248, 1954.

[18] W.H. Hayt and J.A. Buck, *Engineering Electromagnetics*, McGraw Hill, New York, 2006.

[19] L. Katz and A.S. Penfold, Range-energy relations for electrons and the determination of beta-ray end-point energies by absorption, *Reviews of Modern Physics*, vol. 24, 1, pp. 28–44, 1952.

[20] R.F. Egerton, Electron energy-loss spectroscopy in the TEM, *Reports on Progress in Physics*, vol. 72, article #016502, 2009.

[21] Donald Neamen, *An Introduction to Semiconductor Devices*, McGraw-Hill, New York, 2005.

[22] M.R. Wehr, J.A. Richards, and T.W. Adair, *Physics of the Atom*, 4th ed., Addison-Wesley, New York, 1984.

[23] Manfredo P. Do Carmo, *Differential Geometry of Curves and Surfaces*, Prentice-Hall, Englewood Cliffs, NJ, 1976.

[24] M. Spivak, *A Comprehensive Introduction to Differential Geometry*, 2nd ed., Publish or Perish, Inc., Houston, TX, 1979.

[25] R. A. Knief, *Nuclear Energy Technology*, McGraw-Hill, New York, 1981, p. 72.

[26] G.C. Messenger and M.S. Ash, *The Effects of Radiation on Electronic Systems*, 2nd ed., Nostrand Reinhold, New York, 1992.

[27] L.H. Van Vlack, *Elements of Materials Science and Engineering*, 5th ed., Addison-Wesley, New York, 1985.

[28] Baojun Liu, Kevin P. Chen, Nazir P. Kherani, Tome Kosteski, Stefan Costea, Stefan Zukotynski, and Armando B. Antoniazzi, Tritiation of amorphous and crystalline silicon using T_2 gas, *Applied Physics Letters*, vol. 89, article #044104, 2006.

[29] Nazir P. Kherani, Baojun Liu, K. Virk, F. Gaspari, W.T. Shmayda, S. Zukotynski1, and Kevin P. Chen, Hydrogen effusion from tritiated amorphous silicon, *Journal of Applied Physics*, vol. 103, 2, article #024906, 2008.

[30] Zhiqiang Zhang and Xiaoping Liao, A thermocouple-based self-heating RF power sensor with GaAs MMIC-compatible micromachining technology, *IEEE Electron Device Letters*, vol. 33, 4, pp. 606–608, 2012.

Chapter 8

Integrated Sensor/ Actuator Units and Special Purpose Sensors

This last chapter is dedicated to integrated sensor/actuator units and special purpose sensors. New devices that benefited from nanotechnology, such as new icing detectors for aircraft, new microfluidic sensor/actuator units for micro-robots and inkjet printers, etc., are described here.

8.1 Aircraft Icing Detectors

This section introduces a new type of icing detector for aircraft applications. Unlike the well-known icing detectors that are based on optics, the new detector utilizes α particles. An α particle source is attached to the wing of the aircraft, and a detector that mainly depends on a MOSFET transistor is placed a few centimeters away from the source. As the α particles strike the detector, they deposit their positive charges on the gate of the MOSFET (n-channel), and the transistor turns ON. If, however, a layer of ice builds up on the wing and prevents the α particles from reaching the detector, the MOSFET shuts OFF. The new detector is more reliable than optics-based detectors because it is an integral part of the wing and hence cannot miss the formation of ice.

8.1.1 Introduction and Principle of Operation

The formation of ice on an airplane wing during flight can be catastrophic. Investigation of numerous icing-related accidents has shown that ice deposits alter the perfect aerodynamic shape of the wing, which can lead to stall [1–3]. A number of aircraft icing detectors, all based on optical techniques, have recently been introduced in the market [4]. While an optical detector can quickly and reliably detect the formation of ice, it is usually not an integral part of the wing structure. One potential problem with such an approach is that ice may form on the wing but not the detector [5]. It is therefore very desirable to have an icing detector that is an integral part of the wing structure (particularly the surface) in order to avoid the possibility of a false reading from the detector. It is the objective of this section to introduce such a detector.

Figure 8.1 shows the fundamental principle of operation the new icing detector. As shown, a thin α-particle source is attached to or embedded in the surface of the wing. The amount of radioactive isotope that is used in the present application is only slightly more than the amount used in a conventional household smoke detector, and therefore does not constitute an environmental or safety hazard, even if touched directly by an individual. The emitted α particles strike a small metallic electrode that is aerodynamically shaped and positioned a few centimeters above the source, as shown. The positive charge of the α particles is therefore transferred to the electrode upon impact, and the α particles are neutralized (i.e., become neutral helium atoms). The electrode is directly connected to the gate of an n-channel MOSFET transistor, as shown in the figure. The MOSFET is known to be highly sensitive to a charge on the gate and will be in an ON state as long as the α particles continue to reach the electrode. If a layer of ice forms on the wing and covers the α-particle source, the α particles will be stopped by such a layer, will fail to reach the electrode, and hence the MOSFET will turn OFF. This results in a current $I = 0$ through the device, which can be used to trigger an alarm.

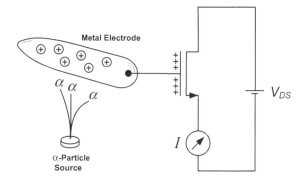

Figure 8.1: Principle of operation of the α-particle icing detector.

A number of important facts must now be mentioned:

■ A layer of water will also stop the α particles in the sensor just described. Hence, the sensor is insensitive to the difference between water and ice. That difference, however, can easily be determined by measuring the temperature of the layer. Hence, the sensor described in Figure 8.1 must be augmented with a thermometer for measuring the temperature of any layer that forms on top of the α-particle source.

■ It is to be pointed out that the presence of clouds or a high degree of moisture in the air does not stop or decrease the range of the α particles [6] (see also the experimental results in this section). Only the presence of a condensed layer of water or ice does stop the α particles. The α particles used in the present application have an energy of 5.5 MeV. Theoretically, these particles can be stopped by a layer of water or ice that is approximately 50 μm thick [7]. It was determined experimentally, however, that a layer that is at least 100 μm (0.1 mm) thick is required to stop the α particles in the present application.[1]

■ The metal electrode shown in Figure 8.1 (which can suitably be an aluminum electrode) can get electrostatically charged as a result of friction with air or dust particles (triboelectric charging). It was determined experimentally, however, that such triboelectric charging will result in a charge that is negligible in comparison with the charge acquired from the alpha particles; and hence will have no effect on the operation of the sensor.

■ Finally, the metal electrode can acquire a substantial amount of spurious charge upon a strike by lightning or from a cloud that holds a significant amount of charge. Fortunately, this effect is transient in nature, and means are incorporated in the circuitry for immediately routing such unusual accumulations of charges to the airframe in order to avoid possible damage to the MOSFET and the associated circuitry. Clearly, therefore, for a signal from the sensor to be taken as a valid indication of the presence of ice, the signal must last continuously for several seconds (i.e., transients must be suppressed/ignored).

It is to be pointed out that the application of detecting icing during flight is challenging because aircraft typically fly in varying atmospheric conditions; however, the technology introduced in this section is the most reliable icing detection technology that the author is aware of (this fact has been confirmed by wind tunnel testing—see the experimental results in Section 8.1.3). During the past few years, a number of new technologies that attempted to overcome the limitations of optical detectors were introduced by different authors [8–10]. Bassey and Simpson [8] proposed using co-planar surfaces on the wing, where the propagation characteristics of an electromagnetic wave that travels between such surfaces will be affected by the

---

[1]This is attributed to the fact that full surface coverage of the α-particle source does not occur in practice if the layer is very thin.

presence of ice. Hua et al. [9] introduced a sensor in which the contraction or expansion of a sensing tube affects the intensity of a magnetic field in a coil that surrounds the tube. Unfortunately, however, in such designs that rely on electromagnetic principles, noise and spurious signals can be a very serious concern. Lin et al. [10] have proposed using a system consisting of digital cameras and image analysis software for visually detecting the presence of a layer of ice. The reliability of such an approach, however, has not been tested or established.

As stated above, extensive wind tunnel testing has shown that the new technology introduced here is highly reliable for the detection of icing under various flying conditions. In the actual sensor (see Figure 8.1), the MOSFET and all the associated electronics are housed inside the metal electrode. Figure 8.2 shows a photograph of the actual prototype that was built and tested by the author. The prototype is mounted on a model airplane with an aluminum fuselage and aluminum wings. It is to be pointed out that icing always occurs on the leading edges of wings, and a typical leading edge is usually totally covered when icing occurs. Hence, by placing the α-particle source on the leading edge of wing, the formation of ice cannot be missed.

In Figure 8.2, the small gray circle that appears under the sensor is the α-particle source. It is a thin foil that contains 10 μCi of the α-particle emitter [241]Am (half-life = 432 years). For the purpose of testing the prototype in a wind tunnel, the foil was attached to the wing with an adhesive; however, in actual implementations a more robust method of mounting must of course be used. As indicated above, the MOSFET transistor and all the associated electronics are housed inside the aerodynamically shaped electrode that is mounted on the wing. The length of the electrode that appears in the figure is 10 cm (this is the length of the electrode that would be used in a full-size aircraft). The electrode, with all the internal electronics, is isolated from the wing with a nonconductive mount.

Figure 8.3 shows the interface circuit that is used in the present prototype. This is a very basic circuit for sensing the state of the MOSFET under various operating

Figure 8.2: Photograph of the detector, mounted on a model airplane with an aluminum fuselage and aluminum wings. The length of the detector is 10 cm.

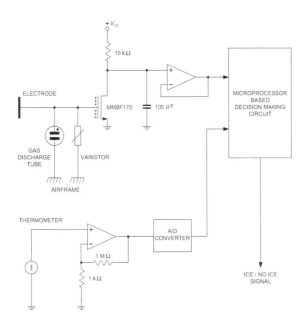

Figure 8.3: Circuit used in the present prototype.

conditions, and more sophisticated circuits can be used in actual implementations, depending on the type of interface that is desired [11,12]. In the circuit, the electrode, which acquires the positive charge, is directly connected to the gate of an n-channel, enhancement mode MOSFET. Two surge protection components are also connected to the gate: a gas discharge tube for protection against lightning, and a varistor for dissipating any voltage build-up above 20 V (the varistor handles a current of up to 100 A, while the gas discharge tube can handle currents of several kA). Both components are directly connected to the airframe, as shown, and help to eliminate transients in addition to the main function of protecting the MOSFET. When the positive charge is present on the gate of the MOSFET, the device is ON, and hence the output voltage (which is fed to a microprocessor circuit) is low. If the charge is not present, however, the device is OFF, and hence the output voltage will be high—indicating an alarm. As shown in the figure, the "alarm" signal is filtered with a low-pass filter for further suppressing the transients. A microprocessor circuit is used for making a decision about the validity of any alarm signal from the MOSFET. The presence of ice on the wing is asserted if two conditions are present:

■ First, the alarm signal (or high-voltage signal) from the MOSFET must hold steady for several seconds; hence any transients are ignored.

■ Second, the thermometer interface circuit, which is shown in the lower part of Figure 8.3, must indicate a below-freezing temperature while the alarm signal from the MOSFET is present.

If the above two conditions are satisfied simultaneously, the microprocessor circuit issues a valid warning signal about the presence of ice on the wing. The microprocessor circuit board that is used in the present application is a small footprint board that is based on an 8-bit PIC microprocessor. The board was programmed in C. As indicated above, all the electronic circuitry shown in Figure 8.3 was included inside the aerodynamically shaped electrode (see Figure 8.2) with the exception of the thermocouple-based thermometer, which was mounted directly next to the α-particle source (the thermometer is too small to be seen in Figure 8.2).

8.1.2 Theory

Determination of the turn ON condition of the MOSFET:

The objective of this section is to determine the conditions that are necessary for turning ON a MOSFET transistor by means of a flow of positively charged particles. The charges that reach the gate of the MOSFET will be used to charge the input capacitance C_{iss} of the device [13]. Depending on the amount of charge, Q, that will be deposited on the gate, the voltage on the gate will reach a final voltage V_{GS} (measured by comparison with the source voltage). The input capacitance is known to be a function of the drain-source voltage, V_{DS} [13]. The deposited charge Q will therefore be given by

$$Q = \int C_{iss}\,dV \tag{8.1}$$

For small MOSFET devices, C_{iss} is usually represented by a straight line with negative slope as a function of V_{DS}. The above integral can be therefore written as follows:

$$Q = \int_{V_{DD}-V_{GS}}^{V_{DD}} (C_0 + mV)\,dV \tag{8.2}$$

where C_0 is the initial capacitance (at $V_{DS} \approx 0$), m is the slope of the straight line, and the limits of integration are due to the fact that the gate-drain voltage changes from V_{DD} to $V_{DD} - V_{GS}$ as the input capacitance is charged (note that the input capacitance is charged while the transistor is OFF, and hence the effective voltage exists between the gate and the drain terminals). By carrying out the simple integration in Eq. (8.2), substituting with the limits, and rearranging the terms, the following quadratic equation is obtained:

$$V_{GS}^2 - 2V_{GS}\left(V_{DD} + \frac{C_0}{m}\right) + \frac{2Q}{m} = 0 \tag{8.3}$$

Solving the above equation for V_{GS} gives

$$V_{GS} = \left(V_{DD} + \frac{C_0}{m}\right) \pm \sqrt{\left(V_{DD} + \frac{C_0}{m}\right)^2 - \frac{2Q}{m}} \tag{8.4}$$

With the values of C_0 and m being available from the datasheet of the device, Eq. (8.4) is an equation in two unknowns: the steady-state charge Q that is present on the gate, and the gate voltage V_{GS}. A second equation that relates these two unknowns can be obtained as follows: under steady-state conditions, the very small current that is supplied by the α particles will constitute a leakage current that flows to the source terminal, through the insulating silicon dioxide (SiO_2) layer that isolates the gate. The leakage current density J through the SiO_2 layer will be related to the electric field intensity E between the gate and the source terminals by the well-known relationship [14]

$$J = \sigma E \tag{8.5}$$

where σ is the conductivity of SiO_2 (this value is approximately $10^{-16}\ \Omega^{-1}m^{-1}$ [15]). The above equation can be written as

$$\frac{I}{A} = \sigma \frac{V_{GS}}{\Delta x} \tag{8.6}$$

where A is the surface area of the gate electrode and Δx is the distance, in general, between the gate and the source (which in practice can be very nonuniform). From Eqs. (8.4) and (8.6), V_{GS} can be expressed as

$$V_{GS} = \frac{I\Delta x}{\sigma A} = \left(V_{DD} + \frac{C_0}{m} \right) \pm \sqrt{ \left(V_{DD} + \frac{C_0}{m} \right)^2 - \frac{2Q}{m} } \tag{8.7}$$

The current I is equal to the ratio $\Delta Q / \Delta t$, where Δt is the time taken to reach the steady state. By replacing Q by ΔQ, the above equation can be written as

$$
\begin{aligned}
V_{GS} &= \frac{\Delta Q}{\Delta t} \left(\frac{\Delta x}{\sigma A} \right) \\
&= \left(V_{DD} + \frac{C_0}{m} \right) \pm \sqrt{ \left(V_{DD} + \frac{C_0}{m} \right)^2 - \frac{2\Delta Q}{m} }
\end{aligned}
\tag{8.8}
$$

By solving the above equation for ΔQ, we obtain

$$\Delta Q = 2 \left(V_{DD} + \frac{C_0}{m} \right) \frac{\sigma A}{\Delta x} \Delta t - \frac{2}{m} \left(\frac{\sigma A}{\Delta x} \Delta t \right)^2 \tag{8.9}$$

Because of the very small value of σ, the second term in the above equation is negligible in comparison with the first term; hence we can conclude that

$$\frac{\Delta Q}{\Delta t} = I \approx 2 \left(V_{DD} + \frac{C_0}{m} \right) \frac{\sigma A}{\Delta x} \tag{8.10}$$

The ratio $\Delta x / A$ can now be estimated by knowledge of the average (typical) value of C_{iss}. According to the datasheet of the MMBF170 transistor used in the present prototype, $C_{iss} \approx 24$ pF [16]. From the well-known equation [14]

$$C = \varepsilon_0 \varepsilon_r \frac{A}{\Delta x} \tag{8.11}$$

where ε_0 is the permittivity of free space, and ε_r is the relative permittivity of the insulating material ($\varepsilon_r = 3.9$ for SiO_2), the ratio $\Delta x/A$ is calculated to be

$$\frac{\Delta x}{A} = \frac{\varepsilon_0 \varepsilon_r}{C} \approx 1.438 \quad \text{m}^{-1} \tag{8.12}$$

Furthermore, the following parameters are given in the datasheet of the device [16]: $C_0 = 40$ pF, $m = -0.5$ pF/V. With $V_{DD} = 6$ V in the present prototype, the current I is calculated from Eq. (8.10) to be approximately equal to 5.8×10^{-15} A. This value will be the "expected" value of the gate current as the gate capacitance charges (the derivation given above was necessary since there is no predefined gate voltage V_{GS}). The question now is whether the current supplied by the α particles will be sufficient to meet this value. As indicated above, the α-particle source used in the present detector contains approximately 10 μCi of ^{241}Am. 10 μCi is equivalent to 370,000 emissions/s. Since the helium nucleus contains two protons, the equivalent of 740,000 positive electron charges will be emitted by the α-particle source each second. Since the α-particle source is a flat film, it is isotropic (i.e., α particles are emitted in all directions). In the present application, it was determined that approximately 10%–20% of the emitted α particles will reach the electrode.[2] The minimum steady-state current that is expected to reach the gate of the MOSFET will be therefore given by

$$I = \frac{\Delta Q}{\Delta t} = \frac{740,000 \times 0.1 \times 1.6 \times 10^{-19}}{1} = 1.18 \times 10^{-14} \quad \text{A} \tag{8.13}$$

Hence, the current available is more than sufficient for charging the gate capacitance and turning the device ON.[3] A direct substitution with this value of I into Eq. (8.7) gives a steady-state voltage V_{GS} that is substantially high. In practice, however, most of the α-particle current was found to leak through the overvoltage protection devices that are present in the circuit (see Figure 8.3), in addition to normal charge loss as the α particles travel through the few centimeters of air that separate the source from the electrode. As a result, the measured steady-state value of V_{GS} was sightly in excess of 10 V. This is clearly well in excess of the threshold voltage of the MOSFET used in the present application (2.1 V according to the datasheet).

The load line and the operating points of the MOSFET:
The load that is attached to the MOSFET in the present application is the 10 kΩ pull-up resistor shown in Figure 8.3. Because the load is large, the MOSFET will be operating in the ohmic region, where the drain current is given by

$$I_D = \frac{V_{DS}}{R_{on}} \tag{8.14}$$

where R_{on} is the ON resistance of the MOSFET. Furthermore, I_D is related to the

---

[2] The collection electrode occupies a solid angle of approximately 70°, which represents 20% of 360°. Accordingly, the collection efficiency is conservatively estimated to be in the range of 10% to 20%.

[3] It is to be pointed out that knowledge of the exact value of the gate capacitance is not necessary, since the supplied gate current is obviously substantially more than the current needed to charge the capacitance.

load by the well-known load line equation [17]

$$I_D = \frac{V_{DD}}{R} - \frac{1}{R}V_{DS} \qquad (8.15)$$

where $R = 10 \text{ k}\Omega$. Equations (8.14) and (8.15) give the value of V_{DS} as

$$V_{DS} = \frac{V_{DD}}{1 + R/R_{on}} \qquad (8.16)$$

With the MOSFET resistance R_{on} being typically equal to 1 Ω, it is clear that $V_{DS} \approx 0$ when the MOSFET is fully ON. Oppositely, however, when the device is fully OFF, $V_{DS} = 6$ V, as can be concluded from Figure 8.3. These are the two operating points of the MOSFET in the present application.

Proof of the foregoing analysis with PSpice simulations and direct measurements:

A PSpice/Multisim simulation of the main part of the circuit in Figure 8.3 is shown in Figure 8.4 below. The objective of the simulation was to prove that the MOSFET will turn ON under the conditions described above, as predicted by the analysis. In the simulation, the α-particle source was replaced with a DC current source with a magnitude of 10^{-14} A. As shown in Figure 8.4, the chosen MOSFET (BS170/MMBF170) does indeed reliably turn ON under the conditions described, as evidenced by the output voltage. It is important to point out that, if the current is reduced to 10^{-15} A, the device still turns ON as predicted by the analysis and confirmed by the simulations.

To further prove the predicted collection efficiency of the electrode, the electrode was connected directly to a small capacitor, and the charge build-up on the capacitor was monitored over a period of several minutes. Since $I = dQ/dt$, the current that is flowing into the electrode can be determined by knowledge of the total collected charge. The current was determined to be indeed in the range of 10^{-15} to 10^{-14} A, as predicted by the analysis in Section 2.1.

8.1.3 Performance Data and Experimental Results

The model airplane to which the detector was attached was tested in a wind tunnel that provides air speeds of up to 1000 km/h (620 mi/h).[4] Figure 8.5 shows a photograph of the first author standing next to the airplane inside the tunnel.

A number of substances were injected into the air flow to test the response of the detector to different air compositions. The following were the compositions/conditions under which the detector was tested:

- Clean, dry air.
- Air with up to 80% relative humidity (moisture).
- Super-saturated water vapor (clouds).
- Air mixed with large, condensed water particles (rain).

---

[4] Access to the testing facilities was provided by ESD Research Inc., a North Carolina corporation.

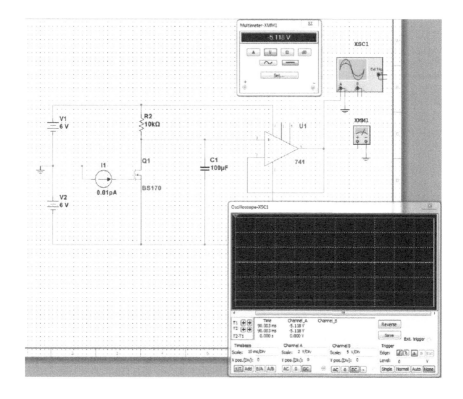

Figure 8.4: PSpice/Multisim simulation of the main part of the circuit in Figure 8.3.

Figure 8.5: Photograph of the author and the model airplane inside the wind tunnel.

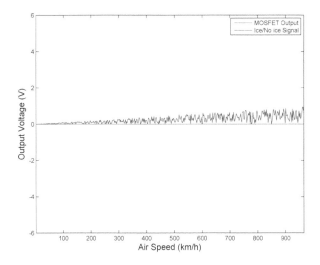

Figure 8.6: Results of testing with dry air, moist air, or super-saturated water vapor: output voltage of the MOSFET, and the ice/no ice signal provided by the decision-making microprocessor, as a function of the air speed.

- Air mixed with small crystals of ice.
- Air mixed with dust particles.
- Artificial lightning.

In the experiments reported here, the following were the atmospheric conditions inside the tunnel:

- Stagnation pressure (or total pressure): approximately 100 kPa.

- Temperature: 15°C (except in one experiment in which ice was injected into the air flow; see Section 3.3).

- Dynamic pressure: the dynamic pressure is given by the well-known quantity $\frac{1}{2}\rho v^2$, where ρ is the density of air and v is the velocity. Clearly this quantity depends on the velocity, as reported in Figures 8.6, 8.7, 8.8, and 8.9.

Results of testing with dry air, moist air, and super-saturated water vapor:
Figure 8.6 shows the results of tests that were conducted with dry air, moist air (80% RH), and super-saturated water vapor (to simulate flying through clouds). The results were very similar for all three conditions, and hence only one set of data is shown in Figure 8.6. The figure shows the output voltage of the MOSFET (measured at the output of the unity gain follower shown in Figure 8.3), together with the ice/no ice signal provided by the decision-making microprocessor circuit. The ice/no ice signal remained at 0 V, as shown, while the MOSFET output remained generally "low" (that is, substantially below 6 V). Hence, the figure shows negative results.

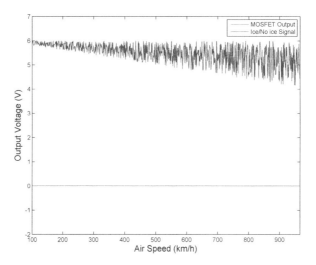

Figure 8.7: Results of testing with simulated rain: output voltage of the MOSFET, and the ice/no ice signal provided by the decision-making microprocessor, as a function of the air speed.

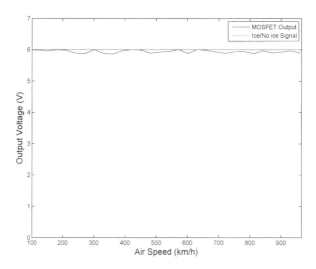

Figure 8.8: Results of testing with ice crystals: output voltage of the MOSFET, and the ice/no ice signal provided by the decision-making microprocessor, as a function of the air speed.

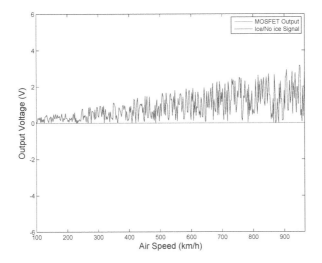

Figure 8.9: Results of testing with dust particles: output voltage of the MOSFET, and the ice/no ice signal provided by the decision-making microprocessor, as a function of the air speed.

The slight increase in the output voltage of the MOSFET at high air speeds is due to the fact that a very small fraction of the α particles are swept away from the detector at such speeds.

Results of testing with large condensed water droplets:

Rain was simulated by injecting water droplets into the air flow (by means of a nozzle that was placed in close proximity to the airplane's wing). Water of sufficient amount was injected to allow the formation of a layer of water on the wing. The results of this test are shown in Figure 8.7. As expected, the output of the MOSFET is high (close to 6 V), as the α-particle source was blocked. However, since the thermometer indicated a temperature that is above $0°C$, the ice/no ice signal remained low (0 V).

Results of testing with small crystals of ice:

This test was performed in the winter, while the temperature outside the tunnel was below freezing (no attempt was made to control the temperature inside the tunnel). Natural snow was injected into the air by using a special pressurized container that was designed by the author's research group. As the snow started accumulating on the wing and covering the α-particle source, measurements were taken. Figure 8.8 shows the results. In this case, both the MOSFET output and the ice/no ice signal were high, indicating a "positive" alarm. It is to be pointed out that a positive signal is generated by the detector only if the thickness of the ice layer exceeds 0.1 mm.

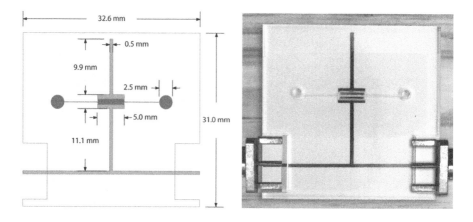

Figure 8.10: (a) Scale-drawn schematic of fabricated microfluidic varactor, from [24]. (b) Fabricated microfluidic varactor with mounted SMA connectors for performing microwave capacitance measurements, from [24].

Results of testing with dust particles:

Airplanes sometimes encounter dust storms. The concentration of dust in such storms typically ranges between 5 and 15 mg per cubic meter of air [18]. The size of the dust particles is typically found to be between several micrometers and 0.1 mm [18, 19]. Dust with such specifications was injected into the air flow in order to test the response of the sensor in the presence of dust. The results are shown in Figure 8.9. As can be concluded, the output of the MOSFET is generally low and intermittent, which is a substantially different response in comparison with the response shown in Figure 8.8. Once again, the microprocessor circuit interpreted the MOSFET's signal as negative.

Testing under lightning strikes:

In this test, lightning was simulated by grounding the airframe and discharging an electrode that was raised to a potential of 2 million volts[5] directly through the detector. As expected, the lightning protection devices included in the detector (see Figure 8.3) quickly dissipated the excess charge within a fraction of a second, and, although the output voltage of the MOSFET swung substantially for a fraction of a second, the transient did not affect the operation of the detector and no false alarm was observed.

8.1.4   Conclusion

The new icing detector introduced in this section is based on highly reliable technology for the detection of ice on aircraft wings and surfaces, as demonstrated by the

---

[5]Such high voltage is typically generated in the laboratory by using a series of diodes and capacitors.

experimental results. Commercially available icing detectors that are based on optical transducers, while technically effective, can be practically unreliable because ice can form on the wing but not on the detector [4,5]. Other icing detection technologies that were very recently described in the literature can suffer from reliability issues as well [8–10]. As described in this section, the new technology is fundamentally based on the generation and the detection of α particles. While the detection methodology introduced here is unique to this application, other techniques for the detection of α particles can be found in the literature [20–23].

8.2 Microfluidic, Microactuators, and Other Special Purpose Small-Scale Devices

The field of integrated sensors/actuators and special purpose sensors is a very advanced field at the present time. Figure 8.10 shows a recently introduced

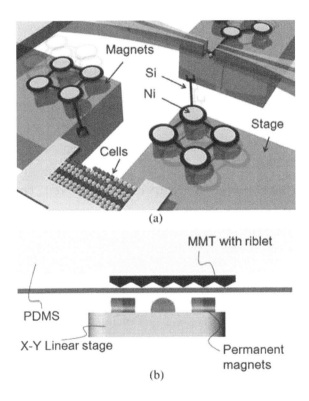

Figure 8.11: Conceptual view of the on-chip cell factory with a high-speed microrobot driven by permanent magnets. The surface of the microrobot has a riblet shape, which is a regularly arrayed V-groove, enabling high-speed actuation of the microrobot, from [25].

inkjet-printed microfluidic RFID-enabled platform for wireless Lab-on-Chip applications (see [24]). The purpose of the sensor is to identify various fluids such as water, alcohol, ethanol, etc.

Figure 8.11 shows a novel, recently introduced concept for a magnetic microrobot actuation in a microfluidic chip

8.3 Quiz

1) True or false: All aircraft icing detectors depend on optics for tracking the formation of ice.
Ans: False. The newest aircraft icing detectors use an α-particle source and an α-particle detector.

2) True or false: Aircraft icing detectors that depend on electromagnetic principles are the most accurate.
Ans: False. These types of detectors are the least accurate.

References

[1] Terry T. Lankford, *Aircraft Icing: A Pilot's Guide*, McGraw-Hill, New York, 1999.

[2] Dennis Newton, *Severe Weather Flying*, Aviation Supplies & Academics, Newcastle, WA, 2002.

[3] Robert N. Buck, *Weather Flying*, McGraw-Hill, New York, 1997.

[4] NewAvionics Corporation, Ice meister model 9732—aerospace ice detecting transducer probe, *NewAvionics Corp, Fort Lauderdale*, FL, http://www.newavionics.com, 2011.

[5] Thomas Meitzler, Darryl Bryk, Euijung Sohn, Mary Bienkowski, Gregory Smith, Kimberly Lane, Rachel Jozwiak, Thomas Moss, Robert Speece, Charles Stevenson, Dennis Gregoris, and James Ragusa, *An Infrared Solution to a National Priority NASA Ice Detection and Measurement Problem*, NASA Report No. 16997, Apr. 2007.

[6] M. G. Holloway and M. Stanley Livingston, Range and specific ionization of alpha-particles, *Physical Review*, vol. 54, pp. 18–37, 1938.

[7] Hervasio G. de Carvalho and Herman Yagoda, The range of alpha-particles in water, *Physical Review*, vol. 88, 2, pp. 273–278, 1952.

[8] C.E. Bassey and G.R. Simpson, Aircraft ice detection using time domain reflectometry with coplanar sensors, *IEEE Aerospace Conference*, pp. 1–6, 2007.

[9] Wang Hua, Huang Xiaodiao, Hong Rongjing, and Fang Chenggang, Establishment and analysis of mathematic model for ice detector, *International Conference on Electronic Measurement and Instruments (ICEMI)*, pp. 714–717, 2007.

[10] Lui Song, Lin Guojianxing, Ma Junqiang, and Wei Hua, The study of ice thickness precise real-time measurement for airplane sensor, *International Conference on MultiMedia and Information Technology (MMIT)*, pp. 618–620, 2008.

[11] W. Boyes, Instrumentation Reference Book, *Butterworh-Heinemann/Elsevier*, Burlington, MA, 2010.

[12] S.A. Dyer, *Survey of Instrumentation and Measurement*, Wiley, New York, 2001.

[13] D. Neamen, *An Introduction to Semiconductor Devices*, McGraw Hill, New York, 2007.

[14] W.H. Hayt and J.A. Buck, *Engineering Electromagnetics*, McGraw Hill, New York, 2006.

[15] James F. Shackelford and William Alexander, *CRC Materials Science and Engineering Handbook*, CRC Press, Boca Raton, FL, 2000.

[16] Fairchild Semiconductor, BS170/MMBF170 N-channel enhancement mode field effect transistor, *Fairchild Semiconductor datasheet*, 2010, www.fairchildsemi.com.

[17] R. C. Jaeger, *Microelectronic Circuit Design*, McGraw Hill, New York, 1997.

[18] Victor R. Squires, *Physics, Mechanics and Processes of Dust and Sandstorms*, Adelaide University, Australia (report), July 2007.

[19] Ilan Koren, Yoram J. Kaufman, Richard Washington, Martin C. Todd, Yinon Rudich, J. Vanderlei Martins, and Daniel Rosenfeld, The bodl depression: a single spot in the sahara that provides most of the mineral dust to the amazon forest, *Environmental Research Letters*, vol. 1, 1, pp. 1–5, 2006.

[20] M. Kurakado, Alpha-particle detection with superconductor detectors, *Journal of Applied Physics*, vol. 55, 8, pp. 3185–3187, 1984.

[21] Toshiyuki Iida, Fuminobu Sato, Yushi Kato, Ippei Ishikawa, and Teruya Tanaka, Development of a simple and tough alpha-particle detector used at high temperature, *Plasma and Fusion Research*, vol. 2, Article ID: S1085, 2007.

[22] G. Charpak, P. Benaben, P. Breuil, and V. Peskov, Detectors for alpha particles and X-rays operating in ambient air in pulse counting mode or/and with gas amplification, *Journal of Instrumentation*, vol. 3, Article ID: P02006, 2008.

[23] L. Rovati, S. Bettarini, M. Bonaiuti, L. Bosisio, G.F. Dalla Betta, V. Tyzhnevyi, G. Verzellesi, and N. Zorzi, Alpha-particle detection based on the BJT detector and simple, IC-based readout electronics, *Journal of Instrumentation*, vol. 4, Article ID: P11010, 2009.

[24] B.S. Cook, J.R. Cooper, and M.M. Tentzeris, An inkjet-printed microfluidic RFID-enabled platform for wireless lab-on-chip applications, *IEEE Trans. on Microwave Theory and Techniques*, vol. 61, 12, pp. 4714–4723, 2013.

[25] M. Hagiwara, T. Kawahara, T. Iijima, and F. Arai, High-speed magnetic micro-robot actuation in a microfluidic chip by a fine V-groove surface, *IEEE Trans. on Robotics*, vol. 29, 2, pp. 363–372, 2013.

Index